AF305020

Savanna Monkeys
The Genus *Chlorocebus*

Living across Africa and the Caribbean, this widely dispersed primate must adapt to different environmental challenges. How do members of the genus *Chlorocebus* live in desert-like conditions and in areas with freezing temperatures and snow in winter? This book examines the ways these primates adapt genetically, hormonally, physically, and behaviorally to their changing landscapes. It features summary chapters for major topics such as behavioral ecology, life history, taxonomy, genetics, and ethnoprimatology. Shorter essays supplement the work, with experts detailing their particular research on these primates. The combination of scholarship provides both a comprehensive view of this adaptable genus while enabling the reader to gain depth in specific topics. Developed from a symposium, this book combines decades of experience working with savanna monkeys into a tangible resource for students and researchers in primatology, as well as for evolutionary and behavioral studies.

TRUDY R. TURNER is Professor of Anthropology at the University of Wisconsin-Milwaukee and an affiliated faculty member in the Genetics Department at the University of the Free State, Bloemfontein, South Africa. She is the co-founder of the Primate Ecology and Genetics Group (PEGG), the Southern African Primatology Society.

CHRISTOPHER A. SCHMITT is Assistant Professor of Anthropology and Biology at Boston University, where he is also affiliated faculty in the Women's, Gender, and Sexuality Program.

JENNIFER DANZY CRAMER is Associate Professor and Program Director of the Sociology, Anthropology, and Women's Studies Program at American Public University System.

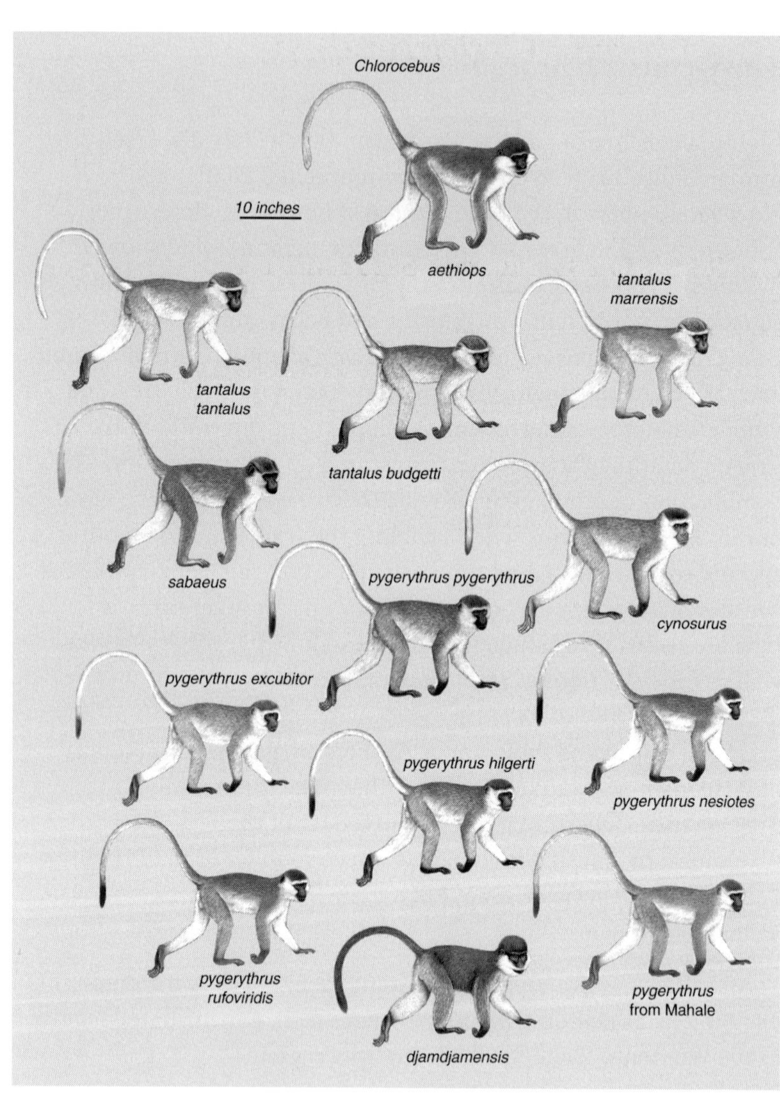

Chlorocebus
10 inches
aethiops
tantalus
marrensis
tantalus
tantalus
tantalus budgetti
sabaeus
pygerythrus pygerythrus
cynosurus
pygerythrus excubitor
pygerythrus hilgerti
pygerythrus nesiotes
pygerythrus
rufoviridis
pygerythrus
from Mahale
djamdjamensis

Savanna Monkeys

The Genus *Chlorocebus*

TRUDY R. TURNER
University of Wisconsin-Milwaukee

CHRISTOPHER A. SCHMITT
Boston University

JENNIFER DANZY CRAMER
American Public University System

CAMBRIDGE
UNIVERSITY PRESS

CAMBRIDGE
UNIVERSITY PRESS

University Printing House, Cambridge CB2 8BS, United Kingdom

One Liberty Plaza, 20th Floor, New York, NY 10006, USA

477 Williamstown Road, Port Melbourne, VIC 3207, Australia

314–321, 3rd Floor, Plot 3, Splendor Forum, Jasola District Centre,
New Delhi – 110025, India

79 Anson Road, #06-04/06, Singapore 079906

Cambridge University Press is part of the University of Cambridge.

It furthers the University's mission by disseminating knowledge in the pursuit of
education, learning, and research at the highest international levels of excellence.

www.cambridge.org
Information on this title: www.cambridge.org/9780521782944
DOI: 10.1017/9781139019941

First published 2019

Printed in the United Kingdom by TJ International Ltd. Padstow Cornwall

A catalogue record for this publication is available from the British Library.

Library of Congress Cataloging-in-Publication Data
Names: Turner, Trudy R., 1950– editor. | Schmitt, Christopher A., editor. |
Cramer, Jennifer Danzy, editor.
Title: Savanna monkeys : the genus Chlorocebus / Trudy R. Turner,
Christopher A. Schmitt, Jennifer Danzy Cramer [editors].
Description: Cambridge, United Kingdom; New York, NY:
Cambridge University Press, 2019. |
Includes bibliographical references and index.
Identifiers: LCCN 2018039944 | ISBN 9780521782944 (hardback)
Subjects: LCSH: Cercopithecus aethiops. | Primates – Behavior.
Classification: LCC QL737.P93 S28 2019 | DDC 599.815–dc23
LC record available at https://lccn.loc.gov/2018039944

ISBN 978-0-521-78294-4 Hardback

In memory of our friend and colleague
Patricia Whitten

Contents

Contributors

Katherine R. Amato, Department of Anthropology, Northwestern University, Evanston, IL, USA

Cristian Apetrei, Center for Vaccine Research, University of Pittsburgh, Pittsburgh, PA, USA; Department of Microbiology and Molecular Genetics, School of Medicine, University of Pittsburgh, Pittsburgh, PA, USA

Louise Barrett, Department of Psychology, University of Lethbridge, Lethbridge, Alberta, Canada; Applied Behavioral Ecology and Ecosystems Research Unit, University of South Africa, Pretoria, South Africa

Maryjka B. Blaszczyk, Department of Anthropology, University of Texas at Austin, Austin, TX, USA

Willem G. Coetzer, Department of Genetics, University of the Free State, Bloemfontein, South Africa

Kerry M. Dore, Department of Anthropology, Baylor University, Waco, TX, USA

Lynn A. Fairbanks, Department of Psychiatry and Biobehavioral Sciences, UCLA, Los Angeles, CA, USA

Nelson B. Freimer, Center for Neurobehavioral Genetics, UCLA, Los Angeles, CA, USA

Tegan J. Gaetano, Department of Anthropology, University of Wisconsin-Milwaukee, Milwaukee, WI, USA

J. Paul Grobler, Department of Genetics, University of the Free State, Bloemfontein, South Africa

Peter Henzi, Department of Psychology, University of Lethbridge, Lethbridge, Alberta, Canada; Applied Behavioral Ecology and Ecosystems Research Unit, University of South Africa, Pretoria, South Africa

Lynne A. Isbell, Department of Anthropology and Animal Behavior Graduate Group, University of California, Davis, Davis, CA, USA

Anna J. Jasinska, Center for Neurobehavioral Genetics, UCLA, Los Angeles, CA, USA; Institute of Bioorganic Chemistry, Polish Academy of Sciences, Poznan, Poland

Joseph G. Lorenz, Department of Anthropology and Museum Studies, Central Washington University, Ellensburg, WA, USA

James E. Loudon, Department of Anthropology, East Carolina University, Greenville, NC, USA

Richard McFarland, Department of Anthropology, University of Wisconsin-Madison, Madison, WI, USA; Brain Function Research Group, School of Physiology, University of Witswatersrand, Johannesburg, South Africa

Michael J. Montague, Department of Neuroscience, Perelman School of Medicine, University of Pennsylvania, Philadelphia, PA, USA

Kimberly Moyer, University of Colorado Denver, Denver, CO, USA

Ivona Pandrea, Center for Vaccine Research, University of Pittsburgh, Pittsburgh, PA, USA; Department of Pathology, School of Medicine, University of Pittsburgh, Pittsburgh, PA, USA

Kevin Raehtz, Center for Vaccine Research, University of Pittsburgh, Pittsburgh, PA, USA; Department of Microbiology and Molecular Genetics, School of Medicine, University of Pittsburgh, Pittsburgh, PA, USA

Rafael L. Rodríguez, Behavioral and Molecular Ecology Group, Department of Biological Sciences, University of Wisconsin-Milwaukee, Milwaukee, WI, USA

Matt Sponheimer, Department of Anthropology, University of Colorado, Boulder, Boulder, CO, USA

Erica van de Waal, Anthropological Institute and Museum, University of Zurich, Zurich, Switzerland; Inkawu Vervet Project, Mawana Game Reserve, KwaZulu-Natal, South Africa

Wesley C. Warren, University of Missouri, Columbia, MO, USA

Patricia Whitten, formerly of Department of Anthropology, Emory University, Atlanta, GA, USA

Brandi T. Wren, Sepela Field Programs, Fort Pierce, FL, USA; Department of Anthropology, Purdue University, Lafayette, IN, USA; Applied Behavioural Ecology & Ecosystem Research Unit, University of South Africa, Johannesburg, South Africa

Preface

Trudy R. Turner, Christopher A. Schmitt,
and Jennifer Danzy Cramer

Savanna monkeys – at times called vervet monkeys and African green monkeys, among many other local names – were always the least studied of the most abundant nonhuman primates. In recent years, this has begun to change. New studies have taken advantage of their wide distribution in different ecological zones to study wide-ranging variation. In addition to behavioral studies on savanna monkeys, over the years, we have extensively sampled animals and obtained morphological, hormonal, and genetic information. We have collected comprehensive information that allows us to take advantage of the most recent technologies to understand the life histories of these animals. It is through the study of these animals in their different environments that we can begin to see the effects of climate and human influence on physiology, morphology, and behavior and elucidate the ways in which evolution operates.

This book was designed to allow the reader to understand savanna monkeys and their place both in the natural world and as a model species for research in several fields. The main chapters of the book describe our own work on savanna monkeys, while separate chapters allow additional researchers to tell the reader about their own work. We want the reader to be aware of the increasing number of people working on savanna monkeys and to have them showcase their own exciting research. The reader can also use this volume to better understand research topics specific to savanna monkeys or to go into more depth on these topics. Our chapter on taxonomy, for example, describes the history of the systematics of savanna monkeys, while additional chapters in this part allow J. Paul Grobler and Willem G. Coetzer to discuss the taxonomic and phylogenetic work on South African vervets in particular, Wesley Warren to talk about new work on the savanna monkey genome, Katherine Amato to present her work on the vervet microbiome, and Cristian Apetrei to discuss how vervets have evolved to respond to taxon-specific infections with simian immunodeficiency virus. We also present basic information on behavioral ecology and historical work on savanna monkeys, while Lynne Isbell discusses vervet social structure, Brandi Wren looks at feeding ecology, Lynn Fairbanks talks about novelty-seeking, and Erica van de Waal discusses cognition. We describe life history theory and how our own work shows the surprising diversity of life history strategies in savanna monkeys, while Rafael Rodríguez discusses sexual selection and morphology and Maryjka Blaszczyk talks about her work on vervet personality. Our final part on ethnoprimatology follows the same pattern – we discuss our own work on savanna monkeys at the human inter-face, while Kerry Dore discusses her work on interactions between savanna monkeys and humans on the island of St. Kitts and James Loudon and Matt Sponheimer discuss the ways in which isotopes can inform what we know about vervet diets. Each part can be read independently, but together, all parts present a complete picture of what we now know about the genus *Chlorocebus*. It is our hope that this will provide readers with flexibility to explore the genus and all of the work that has been done to understand it to date.

This book originally began with the collaboration of Patricia Whitten and Trudy Turner, who met while they were both (young and) conducting fieldwork in Kenya. While Pat went on to explore many other primates, Trudy and her students continued to work primarily with savanna monkeys. Although Pat was not able to work on this volume, we will always remember her joy in working with primates and her insight into the field.

None of our work would have been possible without the help of many people. While we have worked primarily in Ethiopia, Kenya, and South Africa, our work has been as wide-ranging as the genus we study, which has also brought us to Botswana, Zambia, Ghana, The Gambia, and St. Kitts and Nevis. For permission and support in conducting this research, we are indebted to the governments and wildlife offices in all of these countries. We specifically thank the Kenya Department of Wildlife Management; the Gambia Department of Parks and Wildlife Management; Botswana Ministry of Environment and Wildlife and Tourism; Ghana Wildlife Division, Forestry Commission; Zambia Wildlife Authority; Ethiopian Wildlife Conservation Authority; Ministry of Forestry and the Environment, Department of Environmental Affairs, South Africa; Department of Economic Development and Environmental Affairs, Eastern Cape; Department of Tourism, Environmental and Economic Affairs, Free State Province; the Ezemvelo KZN Wildlife in KwaZulu-Natal Province; and the Department of Economic Development, Environment and Tourism, Limpopo Province.

We are also indebted to the many institutions that have hosted us while conducting research, including the Institute of Primate Research, Kenya; the University of Limpopo, South Africa; the University of the Free State, South Africa; the Mammal Research Institute, University of Pretoria, South Africa; Adrian Tordiffe and the Faculty of Veterinary Medicine, University of Pretoria, South Africa; the Medical Research Council (MRC), The Gambia, specifically Martin Antonio, Michel Dione, and Mamkumba Sannch; Gene Redmond of the St. Kitts Biomedical Research Foundation; and all of the wonderful researchers who work as part of the International Vervet Research Consortium.

We thank the International Union for Conservation of Nature (IUCN) Red List for permissions to use *Chlorocebus* distribution maps. We are also grateful to Donna Genzmer and University of Wisconsin-Milwaukee Cartographic Services for map production. We are indebted to Stephen D. Nash/IUCN Primate Specialist Group for the wonderful cover illustration and front plate.

Our research has been conducted over many years and in many places and could not have been done without the many people who have helped us on the ground, in the field, in the lab, and in valuable discussions. For this invaluable assistance, guidance, skilled reading, and support, we thank Cliff Jolly, Fred Anapol, Fred Brett, Nicholas Dracopoli, James Else, Judith Masters, Krista Fish, J. Paul Grobler, Pat

Gray, Adrian Tordiffe, Joe Lorenz, Anna Jasinska, and Nelson Freimer in particular. We have many more people to thank for helping us collect data, including Laura Alport, David Beller, Micah Beller, Sumarie Besler, Magdalena Lukasic-Braun, Annesca Bubb, Maryjka Blaszczyk, Illysees Coetzee, Willem G. Coetzer, Jill Drake, Tegan Gaetano, Angus Hart, Martin Haupt, Peter Heisler, Leizel Herselman, Magali Jaquier, Bea Jordann, Antoinette Kotze, Ute Kryger, Michael Lawes, Thabiso Ledwaba, Hanna Legator, Nicole Lukas, Lizanne Meiring, Steffi Meyer, Adelene Marais, Hannes Marais, Kerry McAuliffe Dore, Oliver "Pess" Morton, Helene de Nys, Alex Nisbett, Thabag Nkadmeng, Pieter Olivier, James Pampush, Brian Peters, Dewald du Plessis, Francoise Rademeyer, Charmaine and Gavin Rous, Marwa Sayed, Claudia Schoene, Patrick Schwab, Murray Stokoe, Nicholas Theron, Elzet Van Aswegen, Bob and Lynn Venter, Rudolf Venter, Wynand Volk, Jonathan Bethard, Morgan Farrar, Christian Gagnon, Stacy-Anne Parke, and Alicia Rich.

Trudy Turner is particularly indebted to her family, Scott, David, and Micah Beller, for their never-ending support.

Christopher Schmitt is, as always, grateful for the support of his family and close friends, especially the patience and (when he's lucky) assistance in the field of his niece, Sasami, and nephews, Kyle and Jaaron.

Jennifer Danzy Cramer thanks her daughters and husband for their patience and encouragement.

Part I Introduction

1 Introduction

Trudy R. Turner, Christopher A. Schmitt,
and Jennifer Danzy Cramer

Savanna monkeys – the ubiquitous, medium-sized animal known as the vervet monkey in East and South Africa, African green monkey in West Africa, and several other local names in other parts of Africa – were one of the first primates studied by contemporary primatologists. The initial studies of savanna monkey behavior took place in East Africa in the 1960s (Jackson & Gartlan, 1965; Gartlan & Brain, 1968; Gartlan, 1969; Struhsaker, 1967a, 1967b, 1967c, 1967d, 1969; Lancaster, 1971; Dunbar & Dunbar, 1974), South Africa (Basckin & Krige, 1973; Krige & Lucas, 1974), and West Africa (Dunbar, 1974) in the 1970s. In the 1980s, these early studies were supplemented by more intensive work on animals at Amboseli and Samburu in Kenya (Cheney et al., 1981; Wrangham, 1981; Cheney & Seyfarth, 1983; Whitten, 1983; Lee, 1984, Isbell et al., 1990), Cameroon (Kavanagh, 1978), Senegal (Harrison, 1983), and South Africa (Henzi & Lucas, 1980; Henzi, 1985). Studies were also conducted on the Caribbean islands of St. Kitts, Nevis, and Barbados, where savanna monkeys have lived for over 300 years after having been transported there from West Africa on ships (Sade & Hildrech, 1965; Poirier, 1972; Horrocks & Hunte, 1983a, 1983b, 1986; Chapman & Fedigan, 1984; Fedigan et al., 1984; Fairbanks & McGuire, 1985; Hunte & Horrocks, 1987; Fedigan & Fedigan, 1988; Boulton et al., 1995). Nonetheless, compared to other widely distributed primate species, like macaques and baboons, savanna monkeys were not so well studied. In the landmark volume, *A Primate Radiation: Evolutionary Biology of the African Guenons* (Gautier-Hion et al., 1988), which both summarized and set the stage for further research, there is only a single chapter devoted to savanna monkeys (Fedigan & Fedigan, 1988). In a follow-up volume, *The Guenons: Diversity and Adaptation in African Monkeys* (Glenn

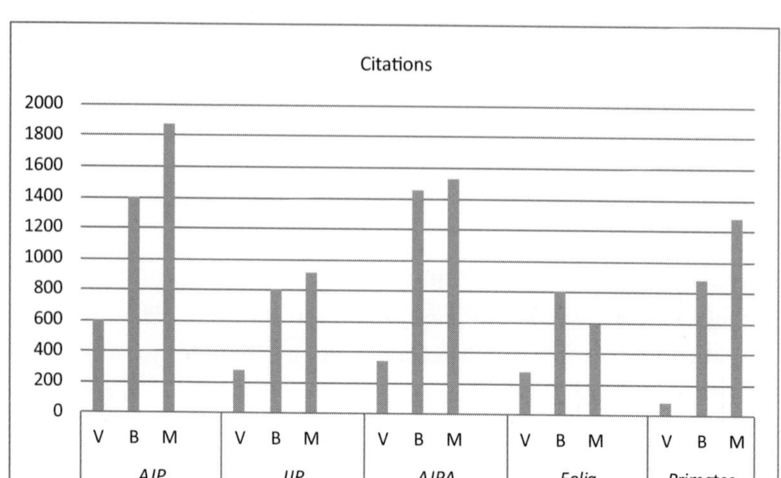

FIGURE 1.1 Number of articles that discuss vervets (V), baboons (B), and macaques (M) in the *American Journal of Primatology* (*AJP*), *International Journal of Primatology* (*IJP*), *American Journal of Physical Anthropology* (*AJPA*) *Folia Primatologica* (*Folia*), and *Primates*.

& Cords, 2002), there is again only a single chapter devoted to savanna monkeys, despite the fact that they are the most numerous and most widespread of all the cercopithecines. An examination of the five major journals that publish work on primates – *American Journal of Primatology*, *International Journal of Primatology*, *American Journal of Physical Anthropology*, *Folia Primatologica*, and *Primates* – clearly documents this lack of focus. Figure 1.1 shows all citations to savanna monkeys, baboons, and macaques in a search of the journals' online content. Savanna monkeys have less than a third of the number of citations compared to baboons and less than a quarter compared to macaques. When viewed over time, the same trend is observed. In Figure 1.2, only articles where the name of the species is found in the title are considered. Again, the number of articles about savanna monkeys is only a fraction of what there is for baboons and macaques.

Why has comparatively little attention been paid to these animals? Some reasons may be that they are smaller in body size than

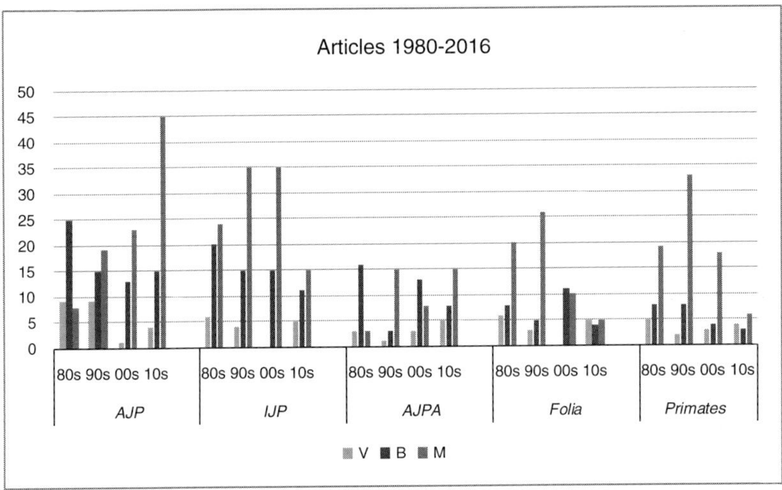

FIGURE 1.2 Number of articles in the decades of the 1980s, 1990s, 2000s, and 2010s that list vervets (V), baboons (B), and macaques (M) in their titles in the *American Journal of Primatology* (*AJP*), *International Journal of Primatology* (*IJP*), *American Journal of Physical Anthropology* (*AJPA*), *Folia Primatologica* (*Folia*), and *Primates*.

baboons and macaques, they have more subtle behavioral signals and smaller group sizes, and they rely more on cryptic behavior to avoid predators. There has been the assumption that they are somehow less interesting than either the forest-dwelling guenons or the larger, savanna-dwelling baboons and macaques. An examination of the articles found in Figure 1.2 may provide some clues – there are a limited number of authors who have published on savanna monkeys, and these articles come from a limited number of sites. Many of those who published on savanna monkeys early in their careers have moved on to other species. Only rarely do you find a researcher continuing at the same field site. This is paired with a downward trend in the number of locations reported on in publications over time – the exact opposite of what has happened for baboons and macaques.

This is beginning to change. In the last few years, more publications from a greater number of locations have begun to emerge. There is a growing recognition that their widespread distribution and concomitant adaptability to a wide array of environments make savanna monkeys a fascinating and useful study subject for a number of questions. Researchers now recognize that savanna monkeys display remarkable differences in group composition (Whitten & Turner, 2004), hormonal profile (Whitten & Turner, 2004), and body type and proportion (Turner, 1997). At the same time, the taxonomy of savanna monkeys has been continually revised, recognizing not only their separation from the rest of the guenons by placing them in a separate genus (Groves, 2001), but also the rich taxonomic variability that can be found within the genus *Chlorocebus.*

Savanna monkeys are widely distributed across sub-Saharan Africa, ranging from the forest/scrubland mosaics of West Africa to the dry, semidesert habitats of northern Kenya and Ethiopia to the snowy heights of the South African Highveld. They are flexible, rapid reproducers, making use of environments as diverse as pristine national parks and reserves and rapidly changing suburban landscapes. The great adaptability of this genus is manifest in these diverse environments through observable changes in body size, social organization, and life history. They are the great opportunists. Although not often used as a model for human evolution, like humans, they appear able to adapt to nearly every environment they encounter. Also like humans, they carry viruses like simian immunodeficiency virus (SIV) – the simian analogue of HIV – that are transmitted sexually and can mutate quickly.

As recognition of savanna monkeys' adaptability has increased, so too has the recognition that much can be learned from this widely distributed group of organisms. It is through the study of differences in localized populations in widespread taxa – where plasticity and adaptive responses to differing environments cause subtle shifts in phenotype – that inferences on the processes of evolution can best be ascertained. The recent resurgence of work on the behavior and biology of vervets (Van de Waal & Bshary, 2011; Pasternak et al., 2013; McFarland et al., 2014; Teichroeb et al., 2015; Henzi et al., 2017), as well as on their genetics and morphology (Turner et al., 2016), is a reflection of this realization.

BRINGING IT TOGETHER: GENETICS, GENOMICS, AND MORPHOLOGY

We have participated in a series of studies on savanna monkeys for several decades (TRT), as well as more recently (JDC and CAS). Over time, our studies have expanded in scope and across the distribution range of the species. The original focus of our work was population genetics, but our work now includes the collection of a wide range of data that are crucial for understanding the life history of savanna monkeys. Our first study of these animals was in Ethiopia in 1973. One of us (TRT) trapped and sampled 125 monkeys from seven troops living in the hot, dry savanna along the Awash River in central Ethiopia (Turner, 1981). Savanna monkeys lived exclusively in the riverine forests and did not venture far from the trees. This study was contemporaneous with the well-known Awash baboon study. Aside from blood for genetic information, the only other biological parameter collected during this study was body weight. Genetic markers were used to determine genetic distance between troops and the rate of migration between troops.

The second study, conducted by TRT, took place in Kenya. As opposed to the close focus of the Awash River study, this Kenya study expanded the scope by sampling at four different sites in Kenya. Each site differed from the others in altitude, rainfall, availability of food, and human presence. While the study was primarily to determine the genetic distances between populations and to track migration between local and more distant groups, greater amounts of biological data were collected on the animals. A series of measures, which we now call "classic" measures, were taken on each animal, along with tooth casts and blood dots to assess parasite infection. We looked at four sites and a total of over 30 groups. Our sampling method in Kenya had two goals: to sample deeply at each site, but also to get a representative sample over the broad distribution of animals in Kenya.

Sites were located between 80 and 300 km apart, and at a single location we sampled anywhere between 5 and 8 groups of animals; each of the groups had between 8 and 25 members. By using this method, we hoped both not to miss any common genetic variants and to really get a sense of the full range of variation in the biological parameters sampled at each location. This strategy of range-extensive and locally intensive sampling has remained our strategy throughout future studies.

An unexpected benefit to one of our sites in Kenya was the presence of a field researcher, Patricia Whitten, who was studying the behavior of vervets in Samburu National Park and agreed to allow us to trap the animals she was observing. While such collaborations may not sound unusual today, it was a radical step forward in the late 1970s. While behavioral ecologists would certainly have been interested in our conclusions, they were not anxious to have their study animals handled in any way, primarily for fear of upsetting their natural behavior. Pat was the exception, and her willingness to allow this led to a three-decade-long collaboration with TRT. Over the course of those decades, new technologies developed that allowed us to examine hormonal variation from serum samples collected from the animals. This led to a multifaceted sample at Samburu, where we had access to behavioral, morphological, hormonal, and genetic data, developing into a rich series of publications that are discussed in this volume.

The third major vervet project of our group began in South Africa as a collaboration between TRT and J. Paul Grobler of the University of the Free State, South Africa. Grobler is a conservation geneticist who was asked by the Department of Environmental Affairs in Limpopo, South Africa, to help solve the vervet "crisis" in the country. Each year, hundreds, perhaps thousands, of vervets are killed in South Africa by cars and people. Vervets are notorious crop and garden raiders and can sometimes make the difference between profit and loss for small farmers. For years, they were considered vermin in South Africa and could be killed at will. But, for as many people who hated vervets, there were those who loved them and wanted to protect and save them. Many of these people set up sanctuaries on their farmland – even in their homes – that would quickly become crowded with injured or orphaned animals. Far from amateurs, these sanctuary owners showed remarkable skill in helping vervets form troops from these unrelated groups animals. Using the vervets' predisposition to allocare, they assigned young, orphaned animals to bond with seasoned mothers, and in that way they slowly integrate into a troop structure – a remarkable feat that has been studied by several research groups (Wimberger et al., 2010; Guy et al., 2011). However, these sanctuaries quickly became overcrowded. Owners wanted to release recovered animals back to the wild and worked hard to find locations where the troops could thrive. At the time, wildlife authorities would not allow releases of animals into areas where they might represent a different subspecies, as defined by some of the older taxonomists. Grobler was enlisted to help solve this dilemma, and TRT joined him as part of a collaboration with the Coriell Institute, which was working to establish a biobank of nonhuman primate tissue samples. Grobler discusses the results of this in this volume. As a result of their work, Grobler and Turner were well positioned to expand their studies.

The Kenya samples were collected at the beginning of what was to become the AIDS epidemic in the early 1980s. Vervets were found to carry SIV, the simian analogue of HIV. Vervets from both Kenya and Ethiopia were found to have the virus, but in very different frequencies (Dracopoli et al., 1983). Vervets became a topic of keen interest as a potential model for HIV, since their own version of the virus appeared to be transmitted sexually. Savanna monkeys had, of course, been used in medical laboratories (where they are usually called African green monkeys) for decades, providing the vector for polio vaccines and as models for human diseases. Their use as an HIV analogue ultimately led to the formation of the fourth major project in which we all (TRT, CAS, and JDC) have participated. This collaborative endeavor ultimately became known as the International

Vervet Research Consortium (IVRC) and has added enormously to our knowledge of vervets across their known range.

Although the IVRC was founded to study the distribution of SIV among savanna monkeys, specifically (see the work of Apetrei and colleagues, this volume), its driving theoretical focus has always been the relationship between genetic/genomic and phenotypic traits, as detailed by Jasinska in this volume. Over the years, the ethics of trapping animals in the wild changed. We have consistently worked on the principle that the trapping and release of animals brings with it the obligation to collect as much information on that animal as possible. In part, it was our goal in this project to ensure that a large-scale trap-and-release study of this kind would not have to be done again. That the tissue we collected could be transformed into immortal cell lines saved in a public repository was part of this goal, providing an ever-ready source of biological information that could be used in conjunction with the phenotypes we worked so hard to collect while trapping.

As a part of the IVRC, we launched major expeditions in South Africa, The Gambia, and St. Kitts and Nevis, along with minor collections in Botswana, Zambia, Ethiopia, and Ghana. We have samples from every taxon of savanna monkey (with the notable exception being the elusive *Chlorocebus djamdjamensis*) and finally have the data to answer some of the major questions about savanna monkeys' adaptations to changing environments and climates. Figure 1.3 presents a comprehensive map of all of the locations we have surveyed. As an expansion of this research, CAS continues trapping and collecting samples from the South African vervet monkeys initially trapped in the IVRC collections in order to gain a better sense of anthropogenic and climate impacts on population processes and evolution in these groups. JDC has continued similar work in The Gambia.

EXPANDING THE SCOPE: RECENT RESEARCH
ON SAVANNA MONKEYS

Of course, we are not the only group conducting research on savanna monkeys. Behavioral studies have proliferated in South Africa over

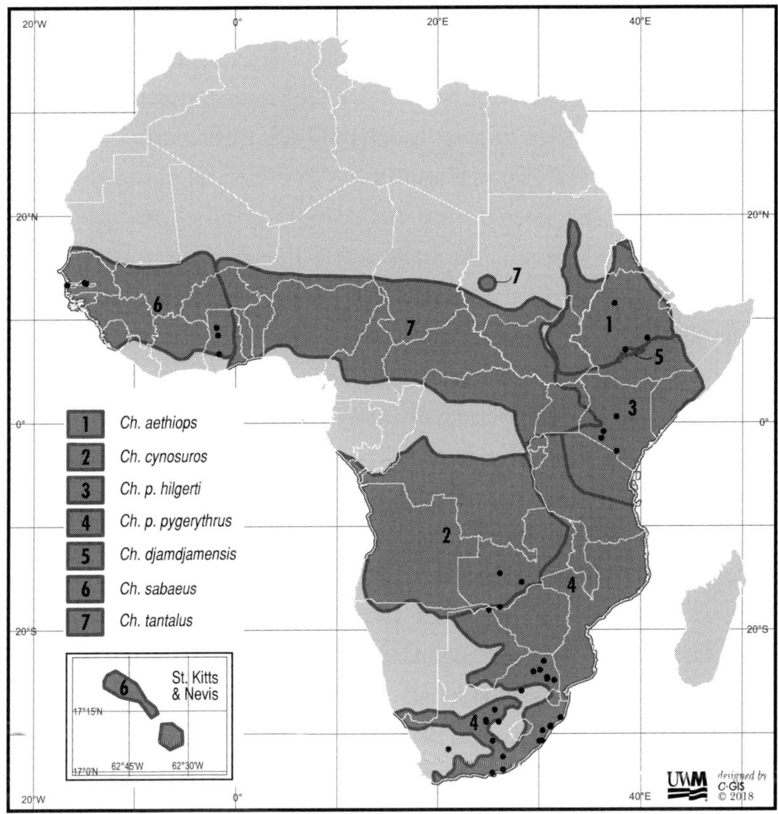

FIGURE 1.3 Location of all savanna monkey trapping sites, 1973–2012. Each dot represents a sampling location. At any single location, between one and nine troops of animals were sampled. This can represent between 8 and 125 animals. Names for animals are taken from Groves (2001).

the past decade. Researchers from South Africa, Europe, Australia, and the USA have begun several short- and long-term research projects. The Applied Behavioral, Ecological and Ecosystem Research Unit (ABEERU) of the University of South Africa (UNISA) is an active and ongoing sponsor of four research sites where students from around the world conduct vervet research (Fruteau et al., 2009; McFarland et al., 2014; Henzi et al., 2016; Wren et al., 2016). Some of these researchers in turn have begun their own projects at their own field sites (Van de Waal et al., 2013). These projects take advantage of the climatic variability in South Africa, as well as the ability of researchers to conduct behavioral and biological experiments on these animals. Additionally, researchers employing new technologies are working on vervets to help understand their life histories. Isotope analysis on hair, variations in gut microbiomes, and field behavioral experiments are all adding to the increasingly complex portrait of this highly adaptable primate. We have asked many of these researchers to describe their work themselves, while we provide the underlying framework and history of research in the genus *Chlorocebus* through the lens of our own work.

We begin by discussing the long history of taxonomy and genetics in savanna monkeys. Here, in particular, there have been enormous changes in the past 15 years. Savanna monkeys were moved into a separate genus, and it is common to now think about the genus as comprising six species. But how confident are we in our definition of separate species? What criteria are we using? How much sense does it make evolutionarily? We provide the background for these discussions, while Grobler and Coetzer discuss their results in studying South African collections, define migration routes into the country, and compare the level of genetic diversity to other primates in the area. Wesley Warren, former director of the McDonnell Genome Institute at Washington University School of Medicine in St. Louis, and Michael Montague have led the efforts to sequence and better understand the structure of the savanna monkey genome. All of the calls for a revision of savanna monkey taxonomy have asked for better genetic information, and that is exactly what Warren and colleagues provide. Apetrei and colleagues further these studies as well by establishing links between SIV's genomic structure and the population genetics and taxonomy of its vervet hosts.

Studies of animal behavior provide the core information for understanding the relationship of the animals with the environment and how this impacts their life history. We review past studies of behavioral ecology, while a number of researchers currently in the field discuss important aspects of behavior uncovered by their own work. Whitten provides a look at animals living in the near desert-like conditions of Samburu in Kenya. Lynne Isbell of the University of California at Davis focuses on predation and food availability as the basic parameters animals must face and describes her long-term research on vervets that led to her limited dispersal hypothesis to explain group living in primates. Brandi Wren discusses her work on vervets in Loskop and the relationship between parasite load and social parameters. Katherine Amato of Northwestern University describes information from gut microbiomes and social relations. Erica van de Waal of the University of Zurich, who also began her work in Loskop but now runs the Inkawu Vervet Project in KwaZulu-Natal, discusses the social learning experiments that have led her and her colleagues to elucidate the methods and strategies vervets use to learn about the environment.

Life history trade-offs are important for understanding both individual and group behavior. Some of the trade-offs are realized through growth and development and others through hormonal regulation. We discuss our own work on growth using a data set of over 2000 animals. This may be the largest set of measures from wild populations for any primate taxon. We are able to compare between taxa and also between populations within taxa from different types of environments. Rafael Rodríguez of the University of Wisconsin-Milwaukee extends this discussion to the static allometry of genital traits. Lynn Fairbanks of the University of California at Los Angeles discusses the relationship between novelty-seeking and the hypothalamic–pituitary–adrenal axis in captive animals. Finally, Maryjka Blaszczyk extends the study of novelty-seeking to observations in the wild by discussing her research on how personality interacts with novelty-seeking behavior in foraging and anti-predator contexts.

Our final chapter discusses new models for understanding primates by focusing on the human–primate interface. These discussions center on ethnoprimatology, a burgeoning discipline that seeks to understand how primates and humans respond to each other at the ever-increasing and permeable boundaries where human settlements and primate habitats meet. Our own work on taxonomy and rehabilitation centers is also discussed. Kerry Dore of the Marist College School of Science presents an analysis of savanna monkeys on St. Kitts, where the "monkey problem" has gotten worse since the end of the sugar economy. James Loudon of East Carolina University and Matt Sponheimer of the University of Colorado at Boulder present a method using stable isotopes to assess the degree of human food that primates eat.

Our ultimate goal with this book is to present the full array of information available for understanding this widely distributed and fascinating primate in order to more ably decipher how a single taxon can adapt to such a wide variety of circumstances. Such an understanding is critical, not only for primate species, but also to serve as a model for understanding how our own species evolved. Savanna monkeys can provide that model.

A note on naming – in this volume, we call all members of the genus *Chlorocebus* "savanna monkeys." The name "savanna monkeys" was first coined by Jonathan Kingdon as a descriptor for this group of animals in the 1970s, although it was not widely used at the time. Not everyone follows this convention. As you will see in the chapter on taxonomy, there is still much confusion about how members of the genus are related to each other. The different groups may be designated as different species or subspecies. They have different vernacular names, some of which are the same as the species or subspecies name – they can be called green monkeys, grivets, tantalus, mahlbrouck, and vervets. Recently, Butyniski, Kingdon, and Kalina (2013) have again referred to all *Chlorocebus* as savanna monkeys; perhaps now this name will gain greater currency. The majority of research on savanna monkeys has focused on what are commonly referred to as vervets. Vervets have an extraordinarily wide range – from eastern through to southern Africa. Here, we use this name when discussing the eastern and southern forms and the more general "savanna monkey" when we are describing other groups or the genus as a whole. Since so much of the work on savanna monkeys has been done on the eastern and southern African vervet, it may almost seem as if the names are used interchangeably. We have tried to adhere to the use of "vervet" for studies that have examined this form and "savanna monkey" for the wider grouping. On occasion, you may note the use of either the taxonomic names used by Groves (2001) or the use of the lowest-level (trivial) scientific name as a vernacular name. In such cases, the name will not be italicized. As a reflection of this variability, contributors to this volume may use other designations. It has been their choice of which name to use.

2 Biological Resources for Genomic Investigation in the Vervet Monkey (*Chlorocebus*)

Anna J. Jasinska

The genetic basis of biological variation has been extensively studied in humans with the focus primarily on traits relevant to human health (Perry et al., 2014; CONVERGE Consortium, 2015; Hibar et al., 2015; Locke et al., 2015; UK10K Consortium et al., 2015). Genetic trait mapping studies, which systematically search across entire genomes (rather than focusing on predefined biologically relevant genes) for loci associated with biological traits have been to a lesser extent conducted in nonhuman primates (NHPs), mainly baboon (Rainwater et al., 2009; Bose et al., 2010; Atkinson et al., 2015; Tung et al., 2015b) and vervet (Freimer et al., 2007; Jasinska et al., 2009, 2012; Jasinska et al., 2017) species. From a biomedical perspective, NHP species represent important models because they are the phylogenetically closest species to humans; they show a high conservation of not only genetic sequence and synteny of genome organization, but also physiological processes, cognition, and behavior, which cannot be effectively modeled in more phylogenetically distant rodent models. From an evolutionary anthropological perspective, genetic studies in various primate species may shed light on the role of genetic variation in the evolution of biological traits in primates and the origins of species-specific traits. Such genetic investigations require species-specific chromosome-level genomic assembly with high-quality gene annotations, as well as large cohorts of phenotypically and genetically characterized individuals.

VERVET AS A MODEL NHP SPECIES

Adequate resources for such investigations have recently been developed for vervets (*Chlorocebus aethiops*), also known as African green monkeys, by the International Vervet/AGM Research Consortium. Vervets have been used for decades as a model in biomedical research and anthropological studies. Vervets have become a major primate model for translational biomedical investigations, particularly in the areas of brain function (Freimer et al., 2007; Woods et al., 2011, Jasinska et al., 2017) and behavior (Fairbanks, 2001; Fairbanks et al., 2001, 2004a, 2004b, 2011, 2012; Laudenslager et al., 2011; Groman et al., 2014), neurodegeneration (Raz et al., 2000; Fainman et al., 2007, Chen et al., 2018), aging (Chichester et al., 2015b), AIDS studies (Chahroudi et al., 2012), diet and cardiometabolic health (Voruganti et al., 2013; Schmitt et al., 2018), obesity (Kavanagh et al., 2007; Bradford et al, 2015), and reproductive biology and female health (Atkins et al., 2014; Chichester et al., 2015a; Kuokkanen et al., 2016). In anthropology and primatology, studies of vervets are predominantly focused on the evolution of primate growth and morphology (Rodriguez et al., 2015), endocrine functions and reproduction (Whitney et al, 2003), and behavior (McFarland et al., 2014). The common denominator of these two perspectives – biomedical and anthropological – is linking genetic variation with phenotypic expression in the context of health, environment, adaptation, and evolution.

Appropriate tools and resources are the foundation for any biological model. The Consortium therefore undertook a large-scale systematic effort in both captive and free-ranging vervet populations to comprehensively characterize genetic variation in parallel with creating a biomaterial bank, phenotypic data collection, and tissue gene expression characterization across vervet subspecies, physical environments, and developmental stages. The Consortium developed advanced genomic tools and data sets, as well as large-scale biomaterial and phenotypic resources from wild and captive populations of vervet monkeys that can now be leveraged in systems genetics studies in biomedical and evolutionary research. Here, we present these advancements of the vervet model for systems-level investigations aimed at integrating multiple levels of biological complexity, from genetic variation through to gene expression regulation, phenotypic variation, commensal and pathogenic microbiomes and viromes, and social and physical environment.

VERVET POPULATIONS

The major strength of the vervet as a model for systems-level studies is the availability of a diversified population for studies in different research settings.

African Vervets: African vervets are among about 40 African NHPs naturally infected with simian immunodeficiency virus (SIV), a close relative of HIV, which causes AIDS pandemics in humans (Sharp et al., 2011). Like other SIV-infected natural SIV hosts, vervets typically lack the clinical signs of development to immunodeficiency. Given the high abundance of African vervet populations and a very high prevalence of SIV among adults, the vervet is a major host reservoir of SIV and a key model species for understanding protective mechanisms against progression to AIDS.

Caribbean Vervets: A small number of vervets from West Africa (van der Kuyl et al., 1995) were introduced to the Lesser Antilles in the Caribbean about 300 years ago (Jasinska et al., 2013). The absence of natural predators and major pathogens allowed a rapid population expansion, reaching the current population of more than 50,000 vervets inhabiting three islands: St. Kitts, Nevis, and Barbados (Jasinska et al., 2009). This severe bottleneck followed by rapid population growth created several advantages for genetic research in Caribbean vervets. First, the limited genetic variation in the Caribbean populations (Warren et al., 2015) simplifies the genetic architecture of complex traits and thus increases the power to identify the genetic loci underlying phenotypic variation. Second, deleterious variants (with significant impact on health) that are present in outbred African populations at low frequencies due to constant elimination by natural selection could have drifted to higher frequencies in the founder Caribbean populations due to rapid demographic expansion. Importantly, the severe host population bottleneck resulted in eradication of two out of three plasma viruses common in African vervets (Kapusinszky et al., 2015), markedly reducing the environmental variation in the plasma virome, one of the most intimate of the body's ecosystem.

In contrast to the widely SIV-infected vervet populations in Africa, Caribbean vervet populations lack the SIV pathogen. The 300-year separation of the SIV-free Caribbean populations from the heavily SIV-infected African populations creates a unique opportunity to observe how the relaxation of selective pressure from the virus has influenced allele frequencies and may provide a natural model for the identification of genetic variants that protect vervets from SIV/AIDS disease. The lack of a major pathogen (SIV) also made Caribbean vervets an attractive model for studies employing experimental SIV infection, as well as a wide range of biomedical studies (see Apetrei and colleagues, this volume).

The Vervet Research Colony (VRC): The VRC is a National Institutes of Health-supported colony of an extended multigenerational pedigree, which expanded from a small number of founders from the Caribbean populations. Specifically, 57 vervets derived from the populations of St. Kitts and Nevis islands were brought to the University of California, Los Angeles (UCLA) about 40 years ago to start the colony. The founders gave rise to more than 2000 descendants over 11 monkey generations, including more than 250 monkeys currently living in the colony. At the beginning of 2008, the colony was relocated to the Wake Forest University Primate Center (WFUPC), where it is currently housed. Establishing the vervet pedigree was therefore preceded by two bottlenecks: a severe bottleneck during the introduction from Africa to the Caribbean and a much milder bottleneck while founding the VRC pedigree. As a result, the VRC broadly reflects the genetic variation of the St. Kitts and Nevis populations (a strong correlation of common alleles exists between the captive pedigree and island population; Jasinska et al., 2012). These two populations comprise a complementary model. The seminatural yet controlled rearing conditions (animals live in breeding groups mimicking those living in the wild) under the controlled environment of the primate facility (including a uniform diet, the absence of major pathogens, and the feasibility of conducting controlled experimental manipulation) considerably reduce environmental variation and thus increase the power of genetic and phenotypic studies. On the other hand, Caribbean populations are highly abundant and thus allow for creating sizable cohorts. In addition, their members are separated by more meiotic steps than pedigree vervets (approximately 90 monkey generations in the Caribbean population versus approximately 11 generations in the VRC; Jasinska et al., 2012), providing the opportunity for higher-resolution and better-powered genetic trait mapping.

INITIAL GENOMIC STUDIES IN VERVETS

Early genomic studies in vervets were focused on the vervets of Caribbean origin as a model for dissecting the genetic basis of complex traits and tissue gene expression relevant to brain and behavior. We focused on three complementary areas of systems biology: genetic regulation of complex traits (such as metabolism of monoaminergic neurotransmitters) by quantitative trait loci (QTL); genetic regulation of steady-state transcript levels by expression QTL (eQTL); and gene interactions with the environment (Jasinska et al., 2013). Despite crude genetic tools, namely the first-generation linkage map comprising 226 microsatellite markers ordered across the autosomal genome (Jasinska et al., 2007), and the lack of a vervet reference genomic sequence, these initial works identified loci for complex traits (e.g., QTL regulating dopamine metabolite level in cerebrospinal fluid [Freimer et al., 2007] and blood eQTL relevant to brain regulation in vervets of Caribbean origin, both in the captive pedigree [Jasinska et al., 2009] and in a natural population [Jasinska et al., 2012]), which demonstrated the potential of the vervet as a model for genetic regulation of brain functions. These initial genetic studies inspired the development of more advanced genomic tools, including a systematic phenotyping effort paralleled with the development of a biomaterial bank.

BIOMATERIAL AND PHENOTYPIC REPOSITORY FROM VERVET POPULATIONS

To facilitate large-scale systems biology studies of the health, development, behavior, and host microbiome in vervets, the Consortium conducted a large-scale trap-release study in vervet populations that created free-ranging cohorts of identifiable monkeys with comprehensive biosamples and life history data. Biological sampling was aimed at creating a resource for biochemical assessments of, primarily but not exclusively, lipid and glucose metabolism, endocrine functions, microbiome composition, host genetic variation, and gene expression. Biological materials were collected from South African *pygerythrus* (>500), West African *Ch. sabaeus* (>150), and St. Kitts and Nevis *Ch. sabaeus* (>950). We banked skin biopsies (to establish fibroblast cell cultures as a long-term source of nucleic acids and cells for in vitro assays), blood, cerebrospinal fluid, hair, and samples of microbiomes populating the gastrointestinal tract, reproductive system, and oronasal cavity (including nasopharyngeal and oropharyngeal samples).

The UCLA Systems Biology Sample Repository (SBSR) handled biological samples with complementary phenotypic characteristics, global positioning coordinates, and subcutaneous microchip information from the more than 1650 wild Caribbean and African vervets sampled between 2009 and 2015. We are leveraging the repository samples to phenotype and genetically characterize diverse vervet populations.

Beyond creating population resources for genetic studies, this repository was developed with a special emphasis on providing materials for studies of natural transmission patterns of microbes and viruses in wild populations in order to shed light on their history and mechanism of spread and the risk of zoonotic transmission to humans. The limited availability of good-quality and sizable biosample sets covering a range of sites (including those separated by major geographic barriers, which can influence the spread of both hosts and microbes) has hampered such studies in NHP populations. The studies of vervets in their natural habitat paralleled with the collection of biological samples have provided a rare insight into viral transmission and pathogenesis in wild populations that would be impossible to obtain with captive monkeys (Ma et al., 2013, Ma, Jasinska et al., 2014; Bailey et al., 2015; Kapusinszky et al., 2015). Importantly, chronically SIV-infected monkeys did not show any significant association of the biomarkers predictive of disease progression and mortality in HIV-infected humans, confirming a benign course of infection in the natural environment. We proposed to leverage these as a natural model to identify natural defense mechanisms against lentiviral infection in vervets that will ultimately help develop a therapy for HIV/AIDS (Ma et al., 2013, Ma, Jasinska et al., 2014; see Apetrei and colleagues, this volume).

NEXT-GENERATION TOOLS FOR GENOMICS STUDIES IN VERVETS

To increase the value of the vervet as a model species, the Consortium focused on developing vervet-specific genomic tools; for example, a reference genomic assembly (Warren et al., 2015), a catalog of genetic variants developed through genome re-sequencing in 163 vervets from African and Caribbean populations (Svardal et al., 2017) and 721 vervets from the VRC pedigree (Huang et al., 2015), a list of transcriptional regulatory elements (Villar et al., 2015), characteristics of developmental and sex-specific patterns of gene expression, and a catalog of tissue eQTL (Jasinska et al., 2017).

Reference Genomic Assembly: The lack of high-quality genomic assemblies from NHPs hampered translational and evolutionary research in species most closely related to humans. Warren et al. created a high-quality genome reference sequence of the vervet (Chlorocebus_sabaeus 1.1) (Warren et al., 2015). It is publically available in two genome browsers: NCBI and Ensembl (see Warren and Montague, this volume). To ensure a high quality of gene model construction and discovery of novel (possibly species- or lineage-specific) genes and transcript isoforms, the Consortium also generated RNA sequencing (RNA-seq) data from multiple internal and peripheral tissues (including brain, neuroendocrine, metabolic, and reproductive tissues) representing various stages of postnatal development.

African vervets evolved under strong selective pressure from viruses, including SIV (Svardal et al., 2017). They developed taxa-specific adaptive responses to SIV, allowing them to avoid AIDS despite SIV infection. Such mechanisms do not act in non-natural hosts like rhesus macaques. With the availability of first two reference genomes from natural hosts of SIV resistant to AIDS, vervets (Warren et al., 2015), and recently also sooty mangabeys (Palesch et al., 2018), a comparative genomic analysis between such natural hosts and vulnerable species (e.g., human or rhesus) becomes a promising approach to revealing genomic loci critical for SIV/HIV infection and the development of AIDS.

Altogether, the construction of the vervet genomic assembly started a new chapter in genomic investigations in the vervet model. The next challenge is to further develop an accurate functional annotation of the genome, in particular identification of various regulatory elements. For example, epigenomic approaches have revealed promoters and enhancers acting in the vervet liver (Villar et al., 2015).

Genome-Wide Genotype Panels for Genetic Mapping Studies from the VRC: The VRC pedigree has been comprehensively characterized with phenotypes related to metabolism, growth, endocrine function, monoamine neurotransmitter metabolism, neuroanatomy, obesity, behavior, cognition, and other traits related to health (Jasinska et al., 2013), as well as to gene expression in blood (Jasinska et al., 2009). To facilitate a discovery of genetic variants contributing to these phenotypes, the Consortium generated high-resolution genetic variation data based on whole-genome sequencing (WGS) in 721 VRC pedigree members (Huang et al., 2015). The subjects in this study were individuals with the most in-depth phenotypic characterization or multiple internal tissues collected during terminal sampling. This effort yielded a catalog of common genetic variants (single-nucleotide polymorphisms [SNPs]) and associated genotype data in each locus. Specifically, two genetic mapping panels were created to suit different genetic analyses: the association mapping SNP set (~500,000 SNPs) and the linkage mapping SNP set (~148,000 SNPs) (Huang et al., 2015). This comprehensive characterization of genome-wide genetic variation in the pedigree will help link rich phenotypic data from sequenced monkeys to specific genomic regions. Given the marked correlation of common genetic variation between the VRC and Caribbean populations (Jasinska et al., 2012), these genetic resources may help in studies undertaken outside the VRC both in Caribbean vervet populations and in other Caribbean-origin vervets.

Developmental Tissue Repository, Tissue Gene Expression, and Compendium of Tissue eQTL: The infeasibility of sampling internal tissues in humans is a major limitation to understanding the links between complex phenotypes (such as behavior or metabolism) and underlying molecular processes operating at cellular and tissue levels. To provide insight into the molecular functions of tissues, the Consortium banked specimens from more than 50 brain regions and more than 20 peripheral tissues (including gastrointestinal, endocrine, and metabolic tissues) from 90 genetically characterized VRC vervets, half males and half females, representing different stages of ontogenic development ranging from neonates to senile, and densely covering early development (i.e., the first year of life). This developmental tissue repository provides a resource for gene expression

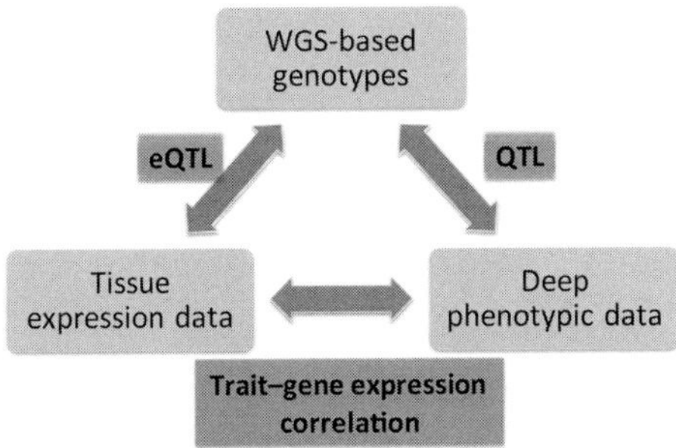

FIGURE 2.1 Supplementing the dissection of QTL for complex traits with gene expression measurements and eQTL discovery in tissues relevant to the trait.

exploration in the context of the gene functions involved in sexual dimorphism (highlighting plausible sex-specific differences in susceptibility to diseases), postnatal development, and genetic regulation of tissue gene expression by eQTL (Jasinska et al., 2017). As trait-associated genetic variants are more likely to be eQTL, the compendium of brain, neuroendocrine, and peripheral tissue eQTL has great potential to improve the dissection of genetic signals for complex phenotypes, which are often linked to broad genetic regions (Figure 2.1). Candidate genes for specific phenotypes can be prioritized based on co-localization within a QTL and correlation of gene expression with trait value. This two-pronged approach revealed lncRNA genes that may participate in a mechanism controlling the size of the hippocampus in vervets (Jasinska et al., 2017).

SUMMARY

The vervet monkey model is an integrated NHP system aggregating a high-quality reference genomic assembly, rich population genomic data collected across a broad geographic range and representing major taxa/populations, a comprehensively phenotyped and genomically characterized pedigree for genetic mapping studies reared under controlled conditions (the VRC), tissue transcriptomic resources, and a compendium of tissue eQTL for high-resolution genetic trait mapping (Figure 2.2). Beyond the studies in the controlled, low-complexity environment of the VRC, the vervet resources were developed for studies in natural populations focused on genetic

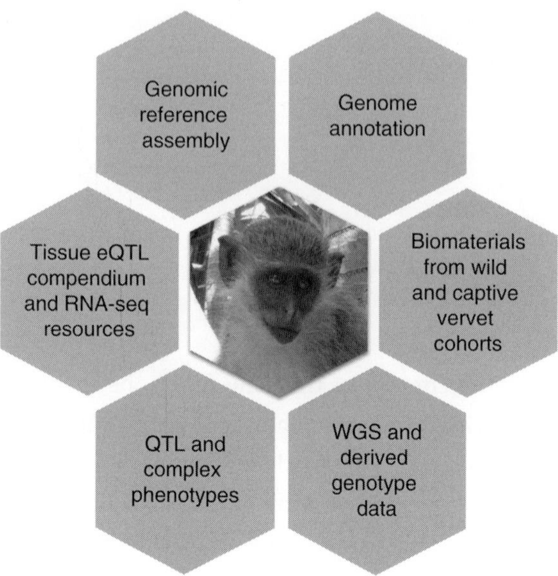

FIGURE 2.2. Systems genomics resources in vervets.

adaptations, phenotype–genotype associations in a context of environmental influences, and the ecology of infectious diseases in natural vervet populations. The creation of the vervet resource by the International Vervet/AGM Research Consortium is a multicenter effort built on the collaboration of many investigators. To maximize the impact of this model, incentivize novel research ideas, stimulate new collaboration, and synergize efforts in different fields, all genomic data presented here were made publically available. The vervet genome is accessible in the NCBI and displayed in the Ensemble genome browser. Genetic variant data were shared via the European Variation Archive at European Molecular Biology Laboratory – The European Bioinformatics Institute (EMBL-EBI). Both WGS and RNA-seq reads data are deposited in NCBI's Sequence Read Archive.

ACKNOWLEDGMENTS

The following grants supported development of the vervet genomic resources described here: 5U54HG00307907, P40RR019963/OD010965, R01RR016300/OD010980, R37 MH060233; UL1-DE019580, R21OD017959 from the US National Institutes of Health, WT095908 and WT098051 from the Wellcome Trust, and aALTF1672-2014 EMBO Advanced Fellowship.

Part II **Taxonomy**

3   Savanna Monkey Taxonomy

Trudy R. Turner, Christopher A. Schmitt,
and Jennifer Danzy Cramer

Every year, hundreds – if not thousands – of savanna monkeys are killed across Africa. Savanna monkeys share habitat with humans and make good use of human food: they raid crops, gardens, garbage, and houses. Humans use poison, guns, and traps to kill these crop raiders, while hundreds of others are killed when they try to cross roads. Yet, for every person who hates these animals, there is another person who loves them and wants to save them. Young orphans are found and rescued. Injured animals are taken for treatment and then allowed to recuperate. In some places, especially South Africa, rehabilitation centers and sanctuaries have been established to care for these displaced animals. The animals are placed into groups and the individuals in the groups learn to live together. These centers soon become crowded and the owners hope to release animals back into the wild – into locations that are being reclaimed as wildlife areas. This, of course, has met with resistance from local farmers and gardeners. Wildlife authorities in South Africa, sensitive to all sides, decided not to allow releases until it could be determined that the releases would not mix animals of different subspecies or evolutionarily meaningful units (EMUs) together. It was only with a clarification of the taxonomy of the animals that releases were allowed to take place. Taxonomy became central to allowing these animals to live in the wild.

SAVANNA MONKEY TAXONOMY

Cercopithecus or *Chlorocebus*? *Chlorocebus aethiops pygerythrus* or *Chlorocebus pygerythrus*? *Cercopithecus aethiops sabaeus* or *Cercopithecus sabaeus* or *Chlorocebus sabaeus*? Vervet monkey? Tantalus monkey? So much confusion. A small monkey of African origin, looking quite similar to other small monkeys of African origin, but with some differences in pelage coloration and whisker shape, widely dispersed across a varying landscape – what are these animals? What difference does it make what we call them – practically? Theoretically? Why is taxonomy important? How do we even define species? Genera?

Taxonomy orders the world and gives a name to a living entity. It is both static and dynamic. It provides a handle by which everyone can recognize what a particular organism might be in a two- or three-word shorthand. But it is also a hypothesis about the relationship between organisms, and such relationships are constantly being reevaluated. New animals are discovered and evaluated; new fossils are unearthed and described; new technologies for evaluating the old order are developed and gain prominence.

Savanna monkey taxonomy is an example of the dynamic process of taxonomy. The vernacular name of this genus is illustrative: only recently has the genus name "savanna monkeys" become popular; historically, the term "vervet" – which specifically refers to the *pygerythrus* grouping in East and South Africa – was often used as a catchall name. The relationships of savanna monkeys to other monkeys in the genus *Cercopithecus*, as well as to patas monkeys, have been historically uncertain as well, and the status and relationships to each of the other major populations are still in flux, and have been for decades. Four questions have emerged as central disputes regarding our understanding of the savanna monkeys' place in nature:

What genus do they belong to – *Cercopithecus* or *Chlorocebus*?
What is the relationship between the rest of the guenons and the savanna monkeys? Is this relationship manifest in the fossil record?
How many species of savanna monkeys are there?
If savanna monkeys are all a single species, exactly how many subspecies are there?

THE GENERIC PLACEMENT OF SAVANNA MONKEYS

Until recently, savanna monkeys were placed in the genus *Cercopithecus*; however, that designation has changed to *Chlorocebus*.

How did that change come about, and what information was used to justify the change? The name *Cercopithecus* derives from the Greek word *kerkopithekos*, which means "tailed monkey." It probably originally referred to the Barbary macaque (*Macaca sylvanus*), which historically had a circum-Mediterranean distribution from North Africa to throughout southern Europe. Other monkeys that reached Greece and Rome in antiquity – with the exception of baboons, which were placed in a separate category – were also probably called *kerkopithecus* or, in Latin, *Cercopithecus* (Groves, 2001). There is documentation (Hill, 1966) that the name *Cercopithecus* was used by Clusii in 1605, more than a century and a half before Linnaeus, for what was possibly a savanna monkey from near the Congo Basin. The name given was *Cercopithecus primus*.

Linnaeus placed all monkeys in *Simia*, and assigned the term *Simia aethiops* for savanna monkeys in 1757 and 1758. *Simia* was soon broken into several families and genera. French scholars did much of this work in the early part of the nineteenth century. It was F. Cuvier who actually named savanna monkeys *Cercopithecus aethiops* in 1819. In 1870, Gray placed savanna monkeys in the genus *Chlorocebus* based primarily on differences in the shape of the head and whiskers from other forest guenons assigned to the genus *Cercopithecus*.

During the 1900s, there were several revisions of *Cercopithecus* taxonomy. The first was by Pocock, the superintendent of the London Zoo from 1904 to 1923 and a researcher at the British Museum, who returned savanna monkeys to *Cercopithecus* and recognized five species. Schwarz, who worked both in biology and medicine, did his major work on primates in the 1920s. In his 1926 evaluation of the cercopithecines, he placed all savanna monkeys into a single species – *aethiops* – in the genus *Cercopithecus*. Groves (2001) considers both Pocock and Schwarz to be great splitters. Pocock split both genera and species, while Schwarz lumped species while splitting subspecies. This is clearly evidenced in their systematic assessment of savanna monkeys. Schwarz also considered all *Cercopithecus* species to be basically identical under the skin. His view of the unity of the genus and the belief that all *Cercopithecus* monkeys – including the forest guenons – were virtually identical under different pelage coloration remained for nearly 80 years. See Table 3.1 for details regarding the changing history of savanna monkey systematics.

Beginning with the introduction of genetic and behavioral techniques for the assessment of relatedness, the generic status of savanna monkeys again came into question. The most recent consensus recognizes a difference between arboreal and terrestrial African Old World monkeys and recognizes that the patas monkey, the l'hoesti monkey group – elevated to the genus *Allochrocebus* (see Tosi et al., 2004; Grubb, 2006; Butynski et al., 2013) – and the savanna monkeys are more closely related to each other than to the rest of the *Cercopithecus* monkeys. This position is based on the accumulated evidence from multiple sources.

Martin and MacLarnon (1988) subjected Verheyen's (1962) cranial and dental measures of a number of cercopithecines to phylogenetic analysis. They characterized morphological differences into Euclidean distances and performed a cluster analysis. The resulting dendrogram clusters *Cercopithecus aethiops* closest to *Allenopithecus* in a grouping including *Miopithecus, Cercopithecus l'hoesti*, and *Erythrocebus*. Strasser and Delson (1987) also linked patas monkeys and savanna monkeys together on the basis of cranio-dental morphology. Groves (1989, p. 135), in his analysis of the Verheyen data, found that patas and savanna monkeys do not have some features found in most other members of the genus *Cercopithecus*. These include: the inferior suborbital region is regularly curved toward the dental arcade; the pyriform aperture is round–oval (not angled in the middle); the temporal lines anteriorly follow the posterior borders of the orbits; the nasal bones usually run straight across inferiorly instead of being pointed in the midline; I^2 is small and pointed and does not bite in a continuation of the occlusal line of I^1; and externally, all of these species have extensive black areas on their limbs. While patas and savanna monkeys lack these derived guenon characters, they share some derived traits not found in the guenons: in side view, the orbits do not slope forward inferiorly, but their lower borders are situated behind the upper margins; the auditory tube has a V-shaped lower margin (a feature often seen, however, in the juvenile guenons); and the orbits themselves are angular instead of oval. Based on these features – both those that are missing from savanna monkeys and patas and present in other cercopiths and those that are present in savanna monkeys and patas and absent from the rest of the cercopiths – Groves placed savanna monkeys in the genus *Chlorocebus*.

Gautier (1988) analyzed vocalizations of the various cerpithecins. *C. aethiops* clusters most closely with *C. preussi* (which is in the *Allochrocebus* or l'hoesti group of guenons) in a group containing patas and talapoin monkeys. Behavioral information has also contributed to an analysis of the taxonomy of savanna monkeys. Most of the rest of the guenons are arboreal, live in rain forests and exhibit a single-male, multi-female social structure. Savanna monkeys, patas monkeys, and talapoins all utilize more terrestrial savanna environments and live in multi-male, multi female groups.

This difference in substrate use is also visible in limb anatomy. In an extensive review of various *Cercopithecus* species, Gebo and Sargis (1994) initially distinguish between arboreal, terrestrial, and semiterrestrial animals. They found that even though savanna monkeys are highly terrestrial and would be expected to group with other terrestrial animals, their limb anatomy was intermediate between arboreal and terrestrial forms, suggesting that terrestriality in cercopithecines evolved multiple times in different lineages and that savanna monkeys evolved this mode independently of patas monkeys. In a later work, however, Sargis et al. (2008) agree with genetic work in asserting that there was only a single event leading to terrestriality in cercopithecins.

Several kinds of genetic data have been used to determine the generic status of savanna monkeys. These data include immunological factors (Goodman & Poulik, 1961), karyotype (Chiarelli, 1968),

Table 3.1 *Designations of vervet taxa through time.*

Vervet taxonomy

Pocock (1907)	Schwarz (1926)	Dandelot (1959)	Hill (1966) superspecies
Cercopithecus aethiops[3]	*Cercopithecus aethiops* (all subspecies of *aethiops*)	*Cercopithecus aethiops* *aethiops* group	*Cercopithecus aethiops*
aethiops	*aethiops*	*aethiops*	*aethiops*
hilgerti[1]	*hilgerti*[1,2]	*hilgerti*[1]	
matschiei[4]		*matschieri*	
ellenbecki			*ellenbecki*
			avattari
Cercopithecus djamdjamensis[5]			
Cercopithecus tantalus			
		tantalus group	
tantalus	*tantalus*	*tantalus*	*tantalus*
budgetti	*centralis*	*budgetti*	*budgetti*
	marrensis	*marrensis*	*marrensis*
Cercopithecus pygerythrus		*Cercopithecus pygerythrus*	*Cercopithecus pygerythrus*
		pyperythrus group	
	arenarius[6]	*arenarius*	*arenarius*
	callidus	*callidus*	*callidus*
centralis		*centralis*	*centralis*
		rubella	

				Location
Dandelot (1971), same as 1959, specific status Napier (1981), all subspecies	Kingdon (1971), one species	Groves (2000). In 1989, *Chlorocebus*, 5 subspecies; in 2000, separate species	Grubb et al. (2003), one species	
Cercopithecus aethiops	*Cercopithecus aethiops*	*Chlorocebus aethiops matschiei –* synonym of *aethiops*	*Cercopithecus* aethiops	
aethiops	*aethiops*		*aethiops*	North Ethiopia/ western border of Nile
hilgerti[1]				Southeast Ethiopia[10]
matschieri				West (lower Omo)[11] West
djamjamensis		*Chlorocebus djamdjamensis*	*djamdjamensis*	
Cercopithecus tantalus		*Chlorocebus tantalus*		
tantalus	*tantalus*	*tantalus*	*tantalus*	Lake Victoria
budgetti	*budgetti*	*budgetti*[9]		Uganda
marrensis		*marrensis* *centralis*		Sudan
Cercopithecus pygerythrus	*pygerythrus*	*Chlorocebus pygerythrus*		
arenarius	*arenarius*			Northern Kenya
callidus	*callidus*			Lake Naivasha
centralis	*centralis*			Lake Victoria, Sessee Island

(*continued*)

Table 3.1 (*cont.*)

Vervet taxonomy

johnstoni	*johnstoni*	*johnstoni*	*johnstoni*
	nesiotes	*nesiotes*	*nesiotes*
rufoviridis	*rufoviridis*[7]	*rufoviridis*	*rufoviridis*
whitei		*whitei*	*whitei*
	excubitor		*excubitor*
		ngamiensis	*ngamiensis*
		marjoriae	*marjoriae*
		helvescens	
	pygerythrus	*pygerythrus*	*pygerythrus*
			cloeti
Cercopthiecus cynosurus		*cynosurus* group	
	cynosurus	*cynosurus*	*cynosurus*
Cercopithecus sabaeus	*sabaeus*	*Cercopithecus sabaeus*	*Cercopithecus sabaeus*
Cercopithecus nigroviridis[8]			

[1] *hilgerti* – Groves moved from *aethiops* to *pygerythrus*.

[2] *hilgerti* – Schwarz replaces *ellenbecki* and *djamdjamensis* with *hilgerti*.

[3] Pocock suggests putting *aethiops* and *tantalus* together and *cynosurus* and *pygerythrus* together.

[4] Possibly *djamdjamensis*.

[5] Very little information besides that it is found at high altitude, possible an offshoot of *aethiops* or *djamdjamensis*.

[6] Replaces older *rufoviridis*.

[7] Replaces *whitei*.

[8] Offshoot of *aethiops* or *pygerythrus*, little information.

[9] Hybridizes with *pygerythrus* around Lake Victoria.

[10] *Djamdjamensis*.

[11] Hybridizes with *pygerythrus*.

				Location
johnstoni	*johnstoni*			Tanzania, south of Kilimanjaro
nesiotes	*nesiotes*	*nesiotes*		Pemba
rufoviridis	*rufoviridis*	*rufoviridis*		East Africa coast/possible central Africa, Zambezi
whytei				East Africa coast
excubitor	*excubitor*	*excubitor*		Manda Island
ngamiensis				Okavango
marjoriae				Botswana
helvescens				
pygerythrus		*pygerythrus*	*pygerythrus*	Great Fish River, South Africa Pilgrim's Rest, South Africa
		hilgerti[1]		
cynosurus	*cynosuros*	*Chlorocebus cynosurus*	*cynosurus*	Kafue/Congo/Namibia
Cercopithecus sabaeus	*sabaeus*	*Chlorocebus sabaeus*	*sabaeus*	

chromosome banding (Ledbetter, 1981; Dutrillaux, 1986; Ponsa et al., 1994), protein data resulting from electrophoresis of sera (Lucotte, 1982; Ruvolo, 1988), mitochondrial genome sequence data (van der Kuyl et al., 1995; Raaum et al., 2005), X and Y chromosome sequence data (Tosi et al., 2002, 2003, 2004, 2005), and next-generation sequencing (NGS) data and targeted mitochondrial genome capture from museum specimens (Guschanki et al., 2013). The earliest genetic technique used to differentiate cercopithecins was immunological distance (see Disotell, 1996 for a fuller explanation), which measures the magnitude of immunological cross-reactivity of proteins from two groups of organisms to assess genetic distance. This and microcomplement fixation, which is similar to immunological distance but uses a reference taxon and measures the antigen–antibody reaction to that for each test group, are both indirect measures of genetic divergence. Studies of chromosome number and banding also began relatively early. While earlier studies were concerned simply with comparing chromosome number, or karyotype, later techniques were devised for staining chromosomes based on the principle that A–T- or G–C-rich areas stain differentially, yielding a chromosome with a distinct pattern of bands. Banding techniques can also highlight deletions, duplications, inversions, and translocations on a chromosomal scale. Any of these may highlight differences in genomic organization between taxa and reveal information about the speciation process. This technique has been particularly important in an assessment of the *Cercopithecus/Chlorocebus* and closely related taxa because of the diversity in terms of chromosome number among species within the genus.

Another early and important technique was protein electrophoresis. The technique was first used in the latter half of the 1960s to determine the amount of genetic variation in natural populations. Data from protein electrophoresis had been successfully used to discuss evolutionary processes, providing estimates of gene flow and migration as well as analyses of genetic distance. The problems with isozyme electrophoresis were effectively solved by the use of DNA sequence analysis. The various techniques look for actual differences in the sequence of bases in DNA between individuals. The first of these techniques did not look at sequences per se, but rather mapped the genome by the locations where bacterial restriction enzymes were able to cut DNA. In restriction mapping, DNA was cut enzymatically into four- or six-base pair fragments, and then the sample was subjected to electrophoresis whereby the patterns and sizes of the fragments were ordered. This provided a rough map of the DNA and was used to estimate strand differences between taxa.

Microsatellite or short tandem repeat (STR) analysis is another technique that has broad applicability. It relies on genomic variation from individual to individual based on the number of sequential (or tandem) repeats of a short sequence of base pairs. There are thousands of these loci in the mammalian genome. Often used to conduct paternity exclusion analyses, in the context of taxonomy STRs are useful, for the most part, for assessing evolutionary processes such as gene flow and genetic drift. They provide data very similar to the data provided by protein electrophoresis in that they provide frequency data for the various alleles that can vary within and across populations. This means that, like electrophoretic data, they can be used to calculate genetic distances between organisms. Using STRs, individuals can be assigned to specific groups, populations, or taxa depending on the frequencies of alleles, from which can be derived likelihood estimates of relatedness and group membership.

A more recent set of techniques involve obtaining the actual sequence of individual base pairs in various kinds of genetic material. For taxonomic assessments, either mitochondrial DNA (mtDNA) or nuclear DNA is used for analysis. mtDNA sequences are smaller in terms of total number of base pairs, but mtDNA is far more numerous in biological samples, as there can be hundreds of mitochondria in a single cell, each with its own copy of the mitochondrial genome, while each cell has only a single copy of nuclear DNA. mtDNA is maternally inherited and, due to mitochondrial involvement in cellular respiration, accumulates mutations and evolves more rapidly than nuclear DNA. Genetic distances for both kinds of DNA are calculated by counting the number of base pair differences between two organisms. There are currently dozens of statistical software packages available to calculate phylogeny based on sequence differences – using a variety of computation algorithms ranging from maximum likelihood to parsimony and encompassing both frequentist and Bayesian modes of inference – that are freely available to users.

Currently, sequencing technology has advanced to the point where we can affordably sequence the whole genome of an organism, enabling us to use distance measures and patterns of similarity between taxa that encompass millions of base pairs as individual character states to determine taxonomy. As these NGS technologies have decreased in price and increased in accessibility, phylogenetic studies based on whole-genome sequencing have increasingly been used to inform taxonomy. Warren and Montague, in this volume, outline how these strategies have been used to tell us more about savanna monkeys in particular. Such work provides the highest level of information yet and not only enables taxonomists to understand the relationships between various taxa, but also can provide reliable estimates of the demographic and evolutionary histories of multiple interbreeding populations that may underlie a species' history. In this way, genomic information of ever-increasing refinement has added to, and in some key ways superseded, traditional methods of taxonomy.

Over time, these different data sets have all combined to add weight to the argument to separate savanna monkeys into a distinct genus more closely allied with patas and l'hoesti monkeys than with the *Cercopithecus* group. Despite some basic differences between their dendrograms, chromosome banding led both Ledbetter and Dutrillaux to link patas monkeys and savanna monkeys closely together. A reanalysis of chromosomal data by Disotell (2000), as well as Ruvolo's (1988) analysis of serum proteins, initially supported this conclusion as well, which was further affirmed by Tosi and colleagues (2002, 2004). Tosi and colleagues (2004) also found that phylogenies derived from one intergenic region of the X chromosome and two Y chromosome genes, *TSPY* and *SRY*, clustered the more terrestrial Cercopithecini together (e.g., the savanna monkeys – *C. aethiops*; patas monkeys, *E. patas*; and the l'hoesti group of monkeys, or *Allochrocebus* – *C. lhoesti*) and separated this larger group from the more arboreal members of the tribe. They suggest that *Allenopithecus* is the basal lineage of the tribe and that this terrestrial grouping argues for there being a single migration from the trees to the ground. This result was affirmed and became consensus by the addition of support for refined chromosomal information (Moulin et al., 2008) and an analysis of *Alu* elements in the nuclear genome of cercopithecins (Xing et al., 2007). Whether these terrestrial groups should ultimately all be lumped into the genus *Chlorocebus* is an issue that has not yet been settled. However, a recent analysis by Guschanski and colleagues (2013), using NGS methods including targeted capture for the mitochondrial genome on a large sample of museum specimens, suggests that when comparing mtDNA data to nuclear data, the split between terrestrial and arboreal species is not as simple as previous analyses suggested, and may actually indicate more than one pathway to the ground.

Some of the most interesting information that bears on the generic placement of savanna monkeys comes from fossils at Maboko Island, Kenya. Benefit and colleagues (1998, 1999, 2001), in a series of publications, have detailed the position of *Victoriapithecus macinnesi*, a fossil dated to the Middle Miocene (or about 15 million years ago), and the only Old World monkey found on the island. Benefit has made the convincing case that these organisms are a sister taxon of the colobines and cercopithecines, or a stem cercopithecoid. It is possible from an examination of *Victoriapithecus* to determine which characteristics are basal for the entire Old World monkey group. These animals are semiterrestrial and most similar to savanna monkeys in their postcranial anatomy (Harrison, 1989). From this, Benefit and Harrison conclude that the use of a semiterrestrial substrate is basal for the Cercopithecoidea. In addition, their dentition suggests that these animals were highly frugivorous, contrary to earlier interpretations that suggested that leaf-eating was the ancestral state for Old World monkeys. Benefit suggests that *Victoriapithecus* exhibited a preference for non-forested habitats and exploited wooded environments with abundant food sources close to the ground. Of particular note is that this information is contrary to ideas that consider arboreal forest-living guenons to be closer to the ancestral form. According to this interpretation, the movement to the ground followed from an arboreal adaptation. *Victoriapithecus*, according to Benefit and colleagues, turns this idea around and argues for semiterrestriality as the original adaptation and that a move back to strict arboreality was secondary. The placement of a savanna monkey-like organism close to the base of the guenon radiation further supports the results of genetic analyses supporting the early divergence of the primarily terrestrial clade to which savanna monkeys and their allies belonged prior to the divergence of the secondarily arboreal forest guenons. There is little other available evidence from the fossil material that sheds light on *Chlorocebus* evolution. While material from other *Cercopithecidae*, particularly *Parapapio*, *Theropithecus* and *Cercopithecoides* from the Pliocene has been found in both eastern and southern Africa, nothing closely resembling modern *Cercopithecus* or *Chlorocebus* is well represented, if at all (Delson, 1984; Frost & Delson, 2002; Jablonski, 2002; Leakey et al., 2003; Frost & Killmer, 2008; Jablonski & Frost, 2010; McKee, van Mayer, & Kuykendall, 2011; Frost, Jablonksi, & Haile-Selassi, 2014; Gilbert, Frost, & Delson, 2016).

SPECIES OR SUBSPECIES?

What is a species? What is a subspecies? Species are the essential building blocks of any taxonomy. Systematics and evolutionary biology recognize the central role of species in understanding the processes and products of evolution. The way that we name and classify species was codified by Linnaeus in the mid-1700s. He provided the structure for the system, which is centered on the binomial system of nomenclature. Every species is given a genus and a species name and is placed into obligatory categories that define its relationship with all other known organisms. The International Commission on Zoological Nomenclature enforces strict and formalized rules for naming taxa, and periodically revises a code that outlines these rules (for a review, see Groves, 2001). While Linnaeus worked a century before Darwin and the beginnings of evolutionary theory, his system provided such flexibility that it has been able to adapt to both a population-based approach to variation in organisms and various novel forms of taxonomic analysis.

In the pre-Darwinian era, species were viewed as divinely created and immutable. In a post-Darwinian, postmodern synthesis view, species are made up of populations that can exhibit variation in multiple traits (Mayr, 1963). This changed view of species, variation, and evolution has led to alternative ways of producing a taxonomy. How do we recognize a species? What about species with wide geographic ranges and multiple subpopulations that differ from each other in only a few traits? How are these populations related to each other? What are the boundaries between these groups and how do these boundaries – or lack of boundaries – inform taxonomy and the processes of evolution?

Ernst Mayr, the great evolutionary biologist, discussed the biological species concept as a way of defining the boundary between organisms. Mayr defined species as "actually or potentially interbreeding populations which are reproductively isolated from other such populations." Species are reproductively isolated by a series of mechanisms that can act either prior to mating or after mating has occurred. How does this definition of a species translate into a taxonomy? In the past, taxonomist often used traits derived from an examination of skins and skeletons, but these traits may not effectively imply reproductive isolation. Reproductive isolation is, in some sense, an "on the ground," fieldwork definition. Taxonomic characters often are not. Groves (1989, p. 2), a leading primate taxonomist, has stated that reproductive isolating mechanisms can be easy to infer from museum specimens:

> If two species can be distinguished by a number of different characters, such that all specimens can at once be referred to one sample or the other, then the existence of two reproductively isolated species can confidently be inferred (provided, of course, that age and sex differences have been taken into account) ... It is even, and quite frequently, valid for a specialist to pick out a single specimen from a larger sample and affirm that it is so different, in so many characters, that it cannot possibly be part of the same species.

But what are the characteristics that are important for distinguishing reproductive isolation, and can these characters be seen in a series of skeletons and skins? While specimens may look different, they are not necessarily different species (the case in point is Groves' placement of savanna monkeys into several different species). And is reproductive isolation still useful as a mechanism to define species?

Due to the difficulty in defining species, there has been a proliferation of species concepts and the properties that define them. While the biological species concept relies on potential interbreeding, the ecological species concept relies on organisms sharing the same ecological or adaptive zone, the recognition species concept requires that organisms share specific-mate recognition systems, the phylogenetic species concept requires that species be either monophyletic or form a diagnosable group, the phenetic species concept requires quantitative differences in multiple traits, and the genotypic species definition requires a genotypic cluster (de Queiroz, 2005). De Queiroz suggests that all of these definitions have in common a single principle – that a species is a metapopulation lineage, or a group of interconnected populations with a unitary gene pool. For de Queiroz, all of the different properties of species that others have used are each aspects of a metapopulation that can be acquired through evolution.

The important point, in this case, is that the metapopulation lineage evolves independently through time.

If we accept de Quieroz's unifying definition of a species, how then do we distinguish it from a subspecies? Is a subspecies simply a local population of a species or a deme? Traditionally, subspecies are seen as clusters of demes within a widespread species that share local adaptations. Miththapala et al. (1996), in their discussion of subspecies and conservation genetics, review the history of the concept:

> Darwin (1859) was well aware of recognizable biotic divisions below the species level, and called these races or incipient species because they appeared to be preludes to species development. Mayr (1963) defined a subspecies as "an aggregate of local populations of a species, inhabiting a geographic subdivision of the range of the species, and differing taxonomically from other populations of the species." Dobzhansky (1937) termed subspecies as "any populations sufficiently distinct to merit a Latin name." More recently Avise and Ball (1990) attempted to provide objective criteria for subspecies recognition by recommending that subspecies designations should be reserved for populations displaying concordant distinctions in multiple, independent, genetically based traits. O'Brien and Mayr (1991) synthesized the subspecies classification to include "individual populations that share a unique geographic range or habitat, a group of phylogenetically concordant phenotypic characters and a unique natural history relative to other subdivisions of the species."

But what are these characters and adaptations? Are they phenotypic differences in whisker length or color adaptations, or are they the result of genetic drift arising from the separation of a small segment of a larger whole? Jolly (1993) suggests that subspecies, like species, arise from evolutionary "events" and, like species, have a unique population history. These taxa are not necessarily incipient species, especially if one accepts a punctuated equilibrium model of rapid speciation. In the classic view, subspecies are considered too ephemeral to be of evolutionary importance because they are only on the way to becoming new species. However, as Jolly (1993, p. 90) states:

> … all taxa are "ephemeral" – all species and subspecies eventually disappear. The question about any taxon is thus not an absolute, Will it disappear? But rather a relative, "How likely is it to persist long enough to provide a foothold for another evolutionary step – another event of divergence?" A measure of security against extinction, both by competition and by accident, is attained only by a small minority of species that reach maturity as comparatively widespread, large-sized and stable populations. However, as the baboons show, subspecies too can be widespread, large and stable.

SO, WHERE DO SAVANNA MONKEYS FIT IN?

Are the differences between local clusters of savanna monkey groups significant enough to be considered species, and if so, how many species are there? If we consider all savanna monkeys as members of a single species, then how many subspecies are we to recognize? The revision of (what then was recognized as) savanna monkey taxonomy in the 1950s and 1960s by Dandelot (1959) and Hill (1966) placed the animals in a superspecies with between three and five species. This splitting also allowed for either 20 or 21 subspecies. These subspecies were defined by a relatively small number of specimens – largely skins and some measurable skeletons with provenance – that existed in museums including the British Museum, the Smithsonian Institution National Museum of Natural History, the Genoa, Paris, Vienna, and Berlin Museums, the Transvaal Museum, and the Museum of Comparative Zoology. Specimens were also available from the Zoological Survey of India. Living animals were also examined at the Washington, London, and Chester Zoos, as well as on the islands of St. Kitts and Nevis, where animals had been transported from West Africa on ships in the 1600s. Subspecific designations were made largely based on the phenetic and morphological characteristics observable on skins, such as presence of a frontal band and pelage coloration (see Table 3.2 for phenotypic distinctions among savanna monkey taxa).

Kingdon (1971) later returned to Schwarz's view of all savanna monkeys as a single species by placing an emphasis on the similarity in forms across taxa and the importance of hybrid forms for taxonomy. He considered there to be five broadly distributed subspecies of savanna monkeys. The range of the subspecies roughly follows geographic zones in Africa. Interestingly, Groves, who first proposed that savanna monkeys were unique enough to be reclassified as *Chlorocebus* (1989), also subsumed all diversity within the genus into a single species with subspecies. In his 2001 work on primate taxonomy, Groves further separated *Chlorocebus* into six species (having elevated the montane Bale monkey of Ethiopia, *Ch. djamdjamensis*, to species level), with numerous subspecies. The 2001 work gives no justification for this change. In fact, the change seems to be contrary to the criterion that Groves states is important for recognizing either species or subspecies, such as allopatric versus sympatric distribution. He states that in order for organisms to be regarded as separate species they should be sympatric, but gives no evidence for this in his designations for *Chlorocebus*. The members of the Orlando Workshop (Grubb, 2003) recognize six subspecies of savanna monkey, also including *djamdjamensis*.

Interestingly, a recent study of the complete mtDNA sequence of the cytochrome b gene of 126 samples of savanna monkeys from across Africa shows little support for any of these designations. Indeed, Haus and colleagues (2013) conclude that, at least for this locus, genetic diversity does not conform to any of the classifications currently in use. Furthermore, in areas of contact between supposed taxa, there was strong evidence of introgressive hybridization as well as considerable mtDNA diversity, which argues that species boundaries – if these different populations are species – are quite permeable. This supports our own work on single-nucleotide

Table 3.2 *Diagnostic criteria for differentiating taxa of vervet monkeys (Turner, 1977).*

Subspecies	Whiskers	Frontal band	Caudal tuft	Feet	Back	Tail tip	Face	Callosities	Scrotum
sabaeus	Yellow[a] semilunar crest around ear	No white band[a]	White with yellow femoral fringe	Pale	Golden green	Golden yellow	Black	Black	Bluish white
tantalus	Brushed up with speckled tip, conceals ear	Separated from whiskers by black temporal bar[a]	White	Pale	Olive green	Creamy white	Black	Black	Sky blue
aethiops	Elongated white whiskers,[a] white moustache	White, straight	White	Pale	Warm olive	Tan	Black	Black	Sky blue
cynosurus	Short whiskers around ear, ear not concealed	Dirty white	Red subcaudal patch[a]	Darkish	Olive gray	Bicolored to tip, tip black	Partially depigmented, flesh colored around eyes[a]	Rose pink[a]	Azure blue

pygerythrus	White bonnet all around face	Pure white, not separated from whiskers	Red subcaudal patch[a]	Black	Gray/ olive	Black	Black	Black	Turquoise blue
djamdjamensis	Bushy white beard	Barely indicated band, separated by black from whiskers	Red/brown tufts	Unknown	Brown	Tufted, reduced tail	Black, mustached	Unknown	Blue

[a] Diagnostic.

polymorphisms (SNPs) and mtDNA. We compared information from Kenya and South Africa. Using subsamples of our original samples, we were able to compare 158 SNPs for 35 animals, 16 from South Africa and 19 from Kenya. Using the same subsample, we compared the 600 base pairs of the mtDNA gene COX1. An unweighted pair group method with arithmetic mean (Michener & Sokal, 1957) tree constructed from these data indicated that the Kenyan animals clustered with more than one taxon and that *pygerythrus* was not monophyletic. This same result was obtained from a STRUCTURE analysis of the SNPs of the same sample (Lorenz et al., 2010; Turner et al., 2010; Grobler et al., 2012).

The most recent work on *Chlorocebus* genomics supports not only novel insights into the population history of the genus – and novel relationships between previously defined subspecies – but also a putative origin of these phylogenetic divisions.

The 2015 publication of the *Chlorocebus* reference genome (based on Caribbean-origin *Chlorocebus sabaeus*, called Chlorocebus_sabaeus 1.1; see Warren and Montague, this volume) was the first step toward enabling researchers to use powerful NGS techniques in order to better understand the population history of these monkeys. With the publication of the reference genome, Warren et al. (2015) discovered divergence dates within *Chlorocebus* that were much more recent than previous estimates, as well as evidence of consistent admixture between taxonomic groups within the genus long considered to be separate species or subspecies. Comparing the reference genome to one individual from each taxonomic group sampled within the genus, they found the most ancient split within *Chlorocebus* to be *sabaeus*, presumably having diverged from a last common ancestor in the genus 531 kya somewhere in Central/East Africa. This was followed by *Ch. aethiops* 446 kya, after which *Ch. aethiops* became genetically isolated from the rest of the genus, presumably by desert-like conditions separating their species range. Central African *Ch. tantalus* is estimated to have diverged from the southern expansion of *Chlorocebus* (comprising *Ch. pygerythrus* and

Ch. cynosuros) 256 kya, followed by the split of *Ch. pygerythrus* and *Ch. cynosuros* at 129 kya.

Svardal et al. (2017) further refined these estimates using 61 million SNPs across the genomes of 163 animals from across their known range and taxonomic groupings, and found a number of surprising results. First, genetic divergence between taxonomic groups was higher than between other primate taxa typically delimited as subspecies (~0.4 percent compared to 0.2–0.32 percent between ape subspecies), but lower than that found between diverged genera (e.g., human and chimpanzee, at ~1.24 percent). Second, despite these relatively deep genetic divergences, population histories among savanna monkeys were surprisingly complex, with high amounts of admixture between geographically adjacent taxa along the east/west gradient of their range, suggestive more of ancient than contemporary gene flow.

Most surprising was the diversity found within the wide-ranging taxon traditionally called *Chlorocebus pygerythrus* (*sensu lato*), which extends from northern Kenya to central and southern Africa along the Indian Ocean coast. Although cluster analyses of sequence data suggest *Ch. hilgerti* (the Kenyan/East African branch of this taxon) to be an outgroup to *Ch. cynosuros* (the Central African taxon) and *Ch. pygerythrus* (the southern African taxon), admixture analyses point to *Ch. cynosuros* being an admixed population of *Ch. hilgerti* and *Ch. pygerythrus* (although the Bostwanan branch of *Ch. cynosuros* appears to have a comparatively large proportion of *Ch. pygerythrus* admixture dated to ~10,000 years before the present, consistent with genetic isolation by distance). Due to this more recent admixture, Svardal et al. (2017) recommend that the taxonomic units of this clade be of a level below those assigned to *Ch. tantalus, Ch. sabaeus,* and *Ch. aethiops.*

Taken together, the phylogenetic findings of Svardal et al. (2017) suggest that the population history of the genus *Chlorocebus* is one of initial admixture between recently diverged populations followed by gradual divergence via isolation by distance. Upon looking for putative signals of selection between populations, however, a further potential driver of divergence was found: diversifying selection for resistance to population-specific strains of simian immunodeficiency virus (SIV). The authors conclude that the coevolution of taxa with their specific strains of SIV should be taken as an important factor driving diversity within *Chlorocebus*. These results are not only relevant for our understanding of the phylogenetic history and taxonomy of the genus *Chlorocebus*, but also point to the savanna monkey as a potentially critical tool for developing translational knowledge and tools for the fight against HIV/AIDS in humans.

The International Vervet Research Consortium genome sequence data are poised to elucidate the relationships between groups of savanna monkeys. To further elaborate on this critical work, in this volume Warren and Montague discuss the process by which genome sequences of animals were obtained and their further utility; Apetrei and colleagues relate the differences between taxa to the SIVagm variants that infect the genus; and Amato discusses the use of savanna monkeys as models for understanding the evolution of the human microbiome.

4 The Promise of Vervet Genomics

Wesley C. Warren and Michael J. Montague

In recent years, we have witnessed the release of numerous non-human primate (NHP) genome reference assemblies (Carbone et al., 2014; Marmoset Genome Sequencing and Analysis Consortium, 2014; Zhou et al., 2014; Warren et al., 2015; Gordon et al., 2016). Each of the represented species either occupy distinct evolutionary lineages within the primate order or serve as prominent biological models for evolutionary and biomedical research. Not surprisingly, most users of such NHP genome models desire human reference assembly quality, mostly to avoid limitations on the hypothesis-driven functional research that arises from computational analyses of primate genomic data. Furthermore, the recent use of genome editing technology to create various models of human disease (Yang et al., 2008; Niu et al., 2014) requires a comprehensive understanding of primate genome structure along with highly accurate gene models.

The recent availability of the high-quality reference assembly for the African green monkey (vervet; *Chlorocebus sabaeus*) represents a first step for enabling numerous types of downstream computational analyses (Warren et al., 2015). The vervet reference assembly was constructed by merging sequence reads from different sequencing technologies (e.g., Illumina, Sanger) with large insert clone resources (e.g., bacterial artificial chromosomes [BACs]). While vervet genome collinearity is generally conserved with human, the vervet karyotype has experienced significant shuffling. For instance, 58 gross synteny breaks were identified overall when compared to human, and this number likely represents an underestimate due to the conservative methods employed (Warren et al., 2015).

At the time of release, the vervet assembly served as the most refined chromosome build compared to other NHP

"

reference assemblies, yet newer sequencing and assembly technologies that incorporate long-read sequence data (e.g., PacBio) have already surpassed the type of hybrid assembly approach used for the vervet reference. The recently published gorilla genome assembly, for example, encompasses near-complete genome representation, with 50 percent of the genome assembled in uninterrupted sequence blocks of 9 Mb (i.e., 9 million bases) or greater, which is surpassed in genome representation and connectivity only by the human assembly (Gordon et al., 2016). Fortunately, plans to replicate this level of quality for the next draft of the vervet genome assembly are now underway. Nonetheless, with the availability of the current vervet reference assembly, initial analyses served to: (1) discern phylogenetic relationships between the known *Chlorocebus* subspecies; (2) identify the ancestral state of variants within bottlenecked populations; and (3) describe functional or even potentially damaging genetic variation within members of the genus.

The subspecies of *Chlorocebus* span diverse habitats throughout their distributions in sub-Saharan Africa. A preliminary characterization of the five taxa revealed a total of 61 million high-quality single-nucleotide variants (SNVs), with 16–27 million SNVs identified per subspecies. This rich genetic diversity is more extreme than previously reported for subspecies of great apes (Prado-Martinez et al., 2013). Using the cumulative catalog of variants within each subspecies, the individual relationships recapitulated earlier phylogenetic hypotheses while producing more refined temporal estimates of subspecies divergence. The genetic signals clearly show an isolation-by-distance pattern within subspecies, with a considerable amount of ongoing gene flow across subspecies (Warren et al., 2015). Further sampling efforts of various African populations will be necessary to understand models of migration and aspects of local adaptation that contribute to the considerable phenotypic diversity within *Chlorocebus*.

In the late seventeenth century, a subset of *Chlorocebus sabaeus* experienced the first of two known bottlenecks, in which populations from West Africa were established on various Caribbean islands. The second bottleneck, from the Caribbean island of St. Kitts to the USA, led to the founding of the Vervet Research Colony (VRC) population (Freimer et al., 2007), established at the University of California, Los Angeles and now maintained at Wake Forest University. The initial comprehensive catalog of SNVs describes the genetic diversity of the VRC population, and with over 720 individuals sampled, the whole-genome sequence data are advantageous for discovering genetic associations with phenotypes of interest (Jasinska et al., 2012). Using the known pedigree structure of the VRC population (Huang et al., 2015), a sequencing strategy for SNV discovery was deployed in which animals located at the higher levels of the pedigree were sequenced to deeper genome coverage (~30×). Genome coverage and, in turn, sequencing cost were subsequently reduced for animals in more recent generations to enable genotype inference and inheritance prediction by imputation. As a result, over 3 million high-quality autosomal SNPs were cataloged for building linkage panels and supporting genome-wide association studies (GWAS) (Huang et al., 2015). As research at the VRC proceeds, characterized SNVs will be of value for augmenting and supporting additional analyses on gene expression patterns and protein function for potentially discovering correlations with phenotypes of relevance for human disease risk, such as obesity (Jasinska et al., 2013; Bradford et al., 2015). The development of this sequencing-based population resource is being replicated using similar sequencing and mapping strategies in other large NHP pedigrees (Haus et al., 2014; Rogers & Gibbs, 2014), such as rhesus macaques (Vinson et al., 2013; Widdig et al., 2015) and baboons (Cox et al., 2013).

Structural variants (SVs) and associated copy number variations are estimated to contribute a much larger share of variation than SNVs (Sudmant et al., 2013). Recent studies have examined the comprehensive scope of structural variation in human genome data while deploying numerous methods that also highlight the technical challenges to uncovering this often overlooked variation (Sudmant et al., 2015: Zarrei et al., 2015). Current NHP genomic resources are limited for understanding heritable and de novo structural variation, likely as a result of fragmented contiguity in the reference assemblies. For example, the lower-contiguity macaque genome resulted in accurate genotyping for only 50 percent of the known SVs in a set of macaque samples (Gokcumen et al., 2013). Likewise, with the current version of the vervet assembly, the existence of notable gaps in contiguity remains a problem for confidently genotyping SVs. For these reasons, the identification of structural variants across NHP species has lagged considerably behind human, and the vervet is no exception (Prado-Martinez et al., 2013), with only the deletion and segmental duplication landscape explored thus far (Warren et al., 2015). Using the whole-genome sequence data from the VRC research population, high-confidence deletion events, measuring 500 bp to 1 Mb in size, were fewer than 200 per individual on average. Additionally, using a small sample from the vervet subspecies, we identified unique deletions, which include a total of 2070 altering exon sequences within 212 genes (Warren, unpublished results). With this small catalog of SVs, efforts are underway to prioritize variants that intersect genic regions with inferred biological relevance, and while computational and technological developments will ultimately improve the sensitivity and accuracy of SV detection, further work is clearly needed to illuminate this important source of genome variation within both captive and wild vervet populations.

In summary, we predict a more complete vervet genome reference assembly in the near future. Accompanying genetic resources, which potentially include a comprehensive, web-based repository for genomic data that displays genome sequences and annotated features (such as physical maps, predicted genes, tandem repeats, expressed sequence tags, single-nucleotide polymorphisms, SVs, etc.), will foster continued population and functional analyses. We envision these resources as being highly integrated with disparate yet complementary types of data, including methylomes, proteomes, and microbiomes. Such a comprehensive resource, created with input and feedback from the end-user community, will serve to fulfill the purposes envisioned by early proponents of vervet genome sequencing, with a long-term goal aimed at improving human health and providing information on primate biology and evolution.

5   African Green Monkeys as a Natural Host of SIV

Cristian Apetrei, Kevin Raehtz,
and Ivona Pandrea

INTRODUCTION

Over 40 different simian immunodeficiency viruses (SIVs) naturally infect African nonhuman primate (NHP) hosts (VandeWoude & Apetrei, 2006; Pandrea, Sodora et al., 2008). SIVs are the root causes of the AIDS pandemic, one of the major public health crises of the twenty-first century (VandeWoude & Apetrei, 2006; Pandrea et al., 2008). Human immunodeficiency virus (HIV) types 1 and 2, the etiologic agents of AIDS, emerged after cross-species transmissions of SIVs from chimpanzees and gorillas in west Central Africa (HIV-1) and from SIVs infecting sooty mangabeys in West Africa (HIV-2) (Hirsch et al., 1989; Gao et al., 1999; Van Heuverswyn et al., 2006).

Each African green monkey (AGM) species was reported to carry a species-specific SIV (type 2). Due to the dispersion and number of AGMs in sub-Saharan Africa, SIVagm represents the largest pool of SIV sequences (Apetrei et al., 2004). Furthermore, because AGMs are not endangered and are readily available for importation from the West Indies (where they are lentivirus-free; Pandrea et al., 2006), AGMs infected with SIVagm are a valuable animal model for the study of nonpathogenic SIV infection in natural hosts.

VIROLOGY

A high genetic diversity is observed within the SIVs naturally infecting AGMs. Each of the four AGM species (vervets, grivets, *tantalus*, and *sabaeus* monkeys), which live in separate geographical regions across Africa, are infected with a species-specific virus (SIVver, SIVgri, SIVtan, and SIVsab), which form monophyletic lineages and cluster together in phylogenetic trees (Fukasawa et al., 1988; Hirsch et al., 1993; Ma et al., 2013). This points to host-dependent evolution and suggests that SIVs were already infecting AGMs at the time of their speciation.

SIVs infecting the different species of AGMs display the common genomic structure of SIVs, consisting of long terminal repeats (LTRs) that flank both ends of the genome, three structural genes (*gag*, *pol*, and *env*), and five accessory genes (*vif*, *vpr*, *tat*, *rev*, and *nef*). They do not carry either of the additional regulatory genes, *vpx* or *vpu* (VandeWoude & Apetrei, 2006).

SIVsab from *sabaeus* monkeys in West Africa is a recombinant virus, with a *gag* gene originating from a Papionini counterpart (most likely SIVrcm from the red-capped mangabey; Jin et al., 1994a; Bailes et al., 2003).

SIVs that naturally infect AGMs display a high propensity for cross-species transmission: in Senegal, West Africa, SIVsab has been transmitted to patas monkeys. In South Africa, SIVver has been transmitted to yellow and chacma baboons (Jin et al., 1994b; Bibollet-Ruche et al., 1996; Van Rensburg et al., 1998).

HOW LONG HAVE SIVS BEEN PRESENT IN NHPS?

The widespread presence of SIVs in numerous African NHPs suggests that SIVs are ancient. However, molecular clock methods indicate a timescale of only centuries to 2000 years (Sharp et al., 2000). In South Africa, the SIVs infecting vervets on the two sides of the Drakensberg Mountains show different clustering patterns. When recalibrating the phylogenetic trees by inferring the time when these two clusters could have been generated – either at the time of AGM spread post-speciation or during the pre-Pleistocene migrations of both of the Bioko sequences – the origin of SIVs could be retraced to 800,000–2,500,000 years ago (Sharp et al., 2000). A low level or absence of pathogenicity of SIVs in AGMs and other natural hosts of SIVs (VandeWoude & Apetrei, 2006) is likely a consequence of long-term host–virus coevolution (Pandrea, Ribeiro et al., 2008; Pandrea & Apetrei, 2010).

The interaction between host and virus on a molecular level can also be used to estimate the extent of primate and lentivirus coexistence. For example, the APOBEC3G region targeted by Vif is adaptively diversified in independent primate lineages, suggesting that the minimum age for the association between Old World monkeys and SIV is around 5–6 million years, and possibly up to 12 million years (Compton & Emerman, 2013).

CHALLENGES TO STUDYING SIV INFECTION IN WILD AGMS

Given the difficulty of accessing primates in their natural habitats, plus the endangered status of several NHP species, noninvasive methods allowing antibody or viral detection in urine or feces have been developed and are currently the strategy of choice for studying SIVs in the wild (Santiago et al., 2002, 2003, 2005). However, such strategies do not permit extensive immunopathogenesis studies. Therefore, the vast majority of data on the natural history of SIVs in their natural hosts was inferred from results obtained in captive monkeys. The only other available data sets collected thus far are from wild AGMs (Ma et al., 2013, Ma, Jasinska et al., 2014).

SIV PREVALENCE AND TRANSMISSION ROUTES IN AGMS

Numerous studies carried out in multiple AGM species from different geographical locations all describe a very high prevalence of SIVs in AGMs. Epidemiologic patterns of seroconversion in wild AGM populations endemically infected with lentiviruses suggest that SIVs are most efficiently transmitted during sexual contact (Ma, Jasinska et al., 2014). Given that HIV is spread by mucosal exposure during sexual contact, it is likely that SIVs are spread via this route as well.

SIV prevalence increases with age, with infants and juveniles having low prevalence, while adult AGMs show extremely high prevalence. SIV prevalence in sexually mature females can reach

80–90 percent, 10-fold higher than in infants and juveniles (Ma et al., 2013). The overall prevalence of SIV in adult males is slightly lower (40–50 percent), but the vast majority of dominant males are SIV-infected. These results point to a very active transmission of SIVs between adult AGMs. The mode of transmission predominating in a given population most likely depends on the behavior and social structure (Phillips-Conroy et al., 1994; Otsyula et al., 1995; Ma et al., 2013).

In addition to sexual transmission, a non-negligible proportion of SIV transmissions in the wild likely occur through biting or fighting, as injuries are frequent and SIV transmission through biting has been described in captive AGMs (Phillips-Conroy et al., 1994). Also, 10 percent of vervets in South Africa showed signs of recent injuries (i.e., deep lacerations) that could result in exposure (Ma et al., 2013). Injuries predominantly occur during the mating season, due to both sexual contact and to contests for dominance between males (VandeWoude & Apetrei, 2006). As monkeys are highly susceptible to SIV infection through oral exposure and most wounds are cleaned through licking (VandeWoude & Apetrei, 2006), the high levels of systemic viral replication documented in the wild (Ma et al., 2013) may facilitate oral transmission during wound cleaning, as suggested by the identification of near-identical virus strains in vervets that were not direct mating partners (Ma et al., 2013).

The rates of maternal-to-infant transmission (MTIT) in AGMs are significantly lower than the 35–40 percent MTIT rates reported in HIV infection (www.unaids.org) or the 40–70 percent MTIT rates in macaques (Amedee et al., 2004), which is likely the result of virus–host coadaptation (Pandrea et al., 2012). Initial studies did not demonstrate MTIT in AGMs (Otsyula et al., 1995), but more recently, we reported cases of infection in infant AGMs at a rate of 4–7 percent, suggesting MTIT as a potential route of SIV transmission in AGMs (Ma et al., 2013, Ma, Jasinska et al., 2014). Note, however, that these rates are below the levels recommended by the World Health Organization as a target for achieving virtual elimination of MTIT, and the exact MTIT route (in utero, perinatally, or via breastfeeding) has not been identified.

NATURAL HISTORY OF SIV INFECTION IN WILD AGMS

In spite of active viral replication and its high prevalence, it is generally assumed that SIV infections are nonpathogenic in African NHP natural hosts, including AGMs (VandeWoude & Apetrei, 2006; Pandrea, Ribeiro et al., 2008; Pandrea & Apetrei, 2010). However, sporadic cases of progression to AIDS in natural hosts were reported (Traina-Dorge et al., 1992; Pandrea et al., 2001; Apetrei et al., 2004; Ling et al., 2004) in monkeys that outlived the life span of their species and had a documented SIV infection for decades (Traina-Dorge et al., 1992; Pandrea et al., 2001; Ling et al., 2004). This supports a paradigm in which SIV infection in African NHPs is a persistent infection with an incubation period that exceeds the normal life span of the natural host (Pandrea et al., 2009). In this paradigm, lack of disease progression is due to active control of the deleterious consequences of SIV infection by the hosts, as the viruses retain their pathogenic potential (Pandrea et al., 2009). Most of the data supporting the lack of pathogenicity of SIVs in natural hosts were derived from the study of only three NHP species in captivity (AGMs, sooty mangabeys, and mandrills; VandeWoude & Apetrei, 2006; Pandrea, Ribeiro et al., 2008; Pandrea & Apetrei, 2010). To draw definitive conclusions about the pathogenicity of SIVs in their natural hosts in the wild, large-scale studies with long-term follow-up are badly needed. Yet, only recently, our large-scale studies in wild AGMs (vervets and *sabaeus* monkeys) reported features of the natural history of SIV infection in the wild (Ma et al., 2013, Ma, Jasinska et al., 2014).

<u>Clinical and Biological Data:</u> None of the wild SIV-infected AGMs tested thus far presented with any of the clinical signs associated with AIDS (i.e., fever, weight loss, lymphadenopathy, or opportunistic infections). The cross-sectional nature of the studies in the wild generally precludes assessment of changes in monkey weight and thus a direct assessment of weight loss. But comparative analyses of the body mass index found that SIV status does not impact the normal weight of wild vervets (Ma et al., 2013).

SIV Receptor Use and Tropism: Most of the SIVs naturally infecting AGMs use the same system of receptors used by HIV-1 (i.e., CD4 as a binding receptor and chemokine coreceptors; VandeWoude & Apetrei, 2006). Unlike HIV-1, which can use both CCR5 and CXCR4 as coreceptors (Moore et al., 2004), most of the SIVs naturally infecting AGMs exclusively use CCR5 (VandeWoude & Apetrei, 2006). SIVsab from *sabaeus* monkeys is an exception, being able to use CXCR4 with no pathologic correlation to this feature (Pandrea et al., 2005). More recently, SIVagm were reported to use the CXCR6 coreceptor (Riddick et al., 2015), similar to SIVs infecting other African NHP hosts (Elliott et al., 2015).

AGMs express lower levels of CCR5 on memory CD4$^+$ T cells compared to progressive hosts (Pandrea, Apetrei et al., 2007). This restriction of CCR5 expression may help: (1) preserve homeostasis of central memory CD4$^+$ T cells despite high levels of viremia (Paiardini et al., 2011); (2) reduce the homing of activated CD4$^+$ T cells to inflamed tissues (Pandrea, Gautam et al., 2007); and (3) moderate SIV transmission (Pandrea, Ribeiro et al., 2008, 2012). Indeed, in wild *sabaeus* monkeys from The Gambia, the levels of CCR5 expression by CD4$^+$ T cells are strongly associated with SIVsab prevalence (Ma, Jasinska et al., 2014).

SIV Replication in Wild AGMs: The levels of chronic viral replication are highly predictive of the outcome of HIV-1 infection (Mellors et al., 1996) and may also predict the outcome of SIV infection in natural hosts, as the NHPs that progressed to AIDS exhibited higher chronic viral loads (VLs) than nonprogressors (Apetrei et al., 2007). To date, VLs in AGMs have been mostly assessed in either experimentally intravenously infected or captive naturally infected monkeys (Diop et al., 2000; Goldstein et al., 2000, 2006; Broussard et al., 2001; Pandrea et al., 2005, 2006, 2012, Pandrea, Ribeiro et al., 2008). Only two studies investigated viral replication in large vervets and *sabaeus* cohorts in the wild (Ma et al., 2013, Ma, Jasinska et al., 2014).

In these studies, SIVver replication levels ranged from 10^4–10^7 SIVver RNA copies/mL (Figure 5.1a), higher in juvenile than in adult

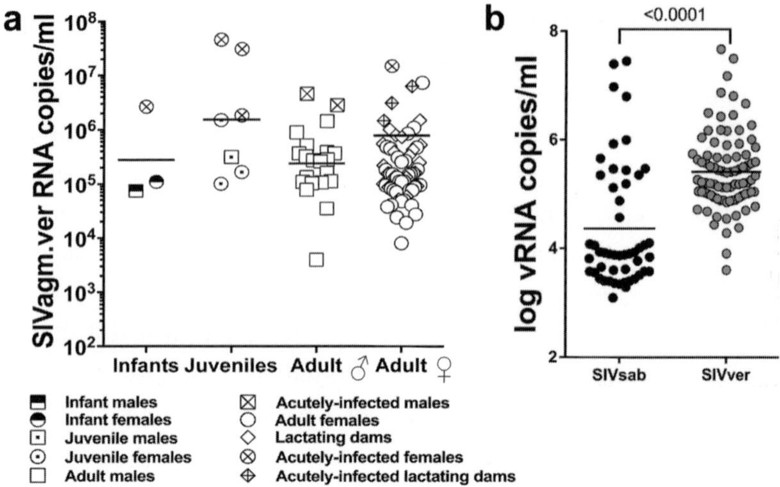

FIGURE 5.1 Plasma VL levels in naturally infected AGMs. (a) VLs in wild SIVver-infected vervets in South Africa revealed no significant difference between infant, juvenile, and adult AGMs or between female and male monkeys. Samples with high VLs assessed as acutely infected are shown. VLs of lactating dams are depicted as diamonds. (b) Comparison between plasma VL in SIV-infected vervets and *sabaeus* monkeys in the wild. Significantly lower levels of plasma VLs were found in naturally infected *sabaeus* monkeys. Detection limit of the real-time polymerase chain reaction assay: 100 copies/mL.

AGMs (Ma et al., 2013). While VLs of the majority of SIVver-infected vervets are in a relatively close range (10^4–10^5 SIVver RNA copies/mL), VLs are higher (10^6–10^8 SIVver RNA copies/mL) in recently infected monkeys (Ma et al., 2013). VLs are slightly lower in wild *sabaeus* monkeys than in vervets (Figure 5.1b), without any pathogenic correlate (Ma et al., 2013, Ma, Jasinska et al., 2014).

Interestingly, lactating dams exhibited high levels of virus (Ma et al., 2013, Ma, Jasinska et al., 2014) (Figure 5.1a), suggesting that the offspring are heavily exposed to SIV during the lactation period. Considering the prevalence of SIV infection in AGM females, there is an enormous in utero exposure of AGM offspring to SIV. However, this massive exposure both in utero and through breastfeeding strongly contrasts with the low SIV prevalence in wild infant AGMs.

The levels of viral replication in wild African NHP hosts of SIVs are in the range of those reported during experimental studies (Pandrea et al., 2006), being one order of magnitude higher than those observed in chronically HIV-1-infected patients (3×10^4 HIV-1 RNA copies/mL; Watkins et al., 2008). Assuming that the shedding of HIV and SIV in the seminal and vaginal fluids of humans and AGMs is similar, the higher chronic SIV VLs in natural hosts are probably significantly greater than the threshold above which the virus can be effectively transmitted through sexual contact (in humans, this threshold of plasma VLs is 1500 copies/mL; Watkins et al., 2008). However, no study has investigated the VLs in seminal fluid in any of the natural hosts of SIVs, including AGMs. Conversely, the high viral replication in wild NHPs and the dramatic increase in SIV prevalence following sexual maturation strongly support sexual contact as the main route of SIV transmission in the wild.

The high levels of chronic viral replication and high prevalence in the wild might also suggest more efficient mucosal transmission of SIV in natural hosts. Yet, the high levels of virus replication are offset by the low CCR5 expression on the mucosal CD4$^+$ T cells. Indeed, characterization of virus diversity in acutely infected AGMs in the wild demonstrated a genetic bottleneck of virus transmission, which was highly similar to those reported in humans and macaques (Keele et al., 2008, 2009), with one to three SIVsab transmitted variants being identified in acutely infected *sabaeus* monkeys from The Gambia (Ma, Jasinska et al., 2014).

Immunopathology of SIV Infection in Wild AGMs: A key factor in the pathogenesis of HIV-induced immunodeficiency is the failure of the lymphoid regenerative capacity (Brenchley et al., 2006; Grossman et al., 2006; Okoye et al., 2007). In particular, bone marrow suppression, reduced thymic output and loss of naive T cells have all been observed in SIV and HIV infections (Roederer, 1995; Douek et al., 1998, 2001; Hellerstein et al., 2003). In addition to the loss of CD4$^+$ T cells, which is the hallmark of HIV/AIDS, an overall immune dysfunction occurs in HIV infection, with alterations of the effectors of adaptive immune response as well as that of innate immune effectors (B cells, dendritic cells, natural killer cells, or monocytes). Altogether, this pan-dysfunction of the immune effectors is critical for driving disease progression to AIDS.

In the only published study on the immunophenotypic changes that occur in naturally SIV-infected African NHPs in the wild, we did not observe any significant changes postinfection with regard to the frequency of CD4[+] T cells, CD8[+] T cells, B cells, monocytes, natural killer cells, or myeloid or plasmacytoid dendritic cells of SIVsab-infected *sabaeus* monkeys from The Gambia (Figure 5.2) (Ma, Jasinska et al., 2014).

The principal reason for the lack of disease progression observed in African NHP hosts appears to be their ability to resolve chronic immune activation at the transition from acute to chronic SIV infection (Pandrea, Apetrei et al., 2007; Estes et al., 2008; Bosinger et al., 2009; Jacquelin et al., 2009; Harris et al., 2010). AGMs maintain the integrity of the mucosal barrier and thus control microbial translocation (MT) (Gordon et al., 2007; Pandrea, Gautam et al., 2007), the main factor behind increased immune activation that drives disease progression in pathogenic HIV/SIV infections (Brenchley et al., 2006; Brenchley & Douek, 2012). It is therefore critical to assess the integrity of the mucosal barrier of AGMs in the wild, where monkeys are exposed to many pathogens in a significantly more hostile environment than captive AGMs. However, a direct assessment of the mucosal barrier in the wild is virtually impossible. We tested the levels of sCD14 in wild AGMs as a surrogate biomarker of MT and showed them to be unchanged in SIV-infected and -uninfected vervets and *sabaeus* monkeys. Even acutely infected monkeys did not show a significant increase in sCD14, in agreement with previous experimental results (Pandrea, Apetrei et al., 2007, Pandrea, Ribeiro et al., 2008), suggesting that wild SIV-infected AGMs maintain the integrity of their mucosal barrier.

Assessment of the levels of chronic immune activation in wild African NHPs is equally difficult and data are only available

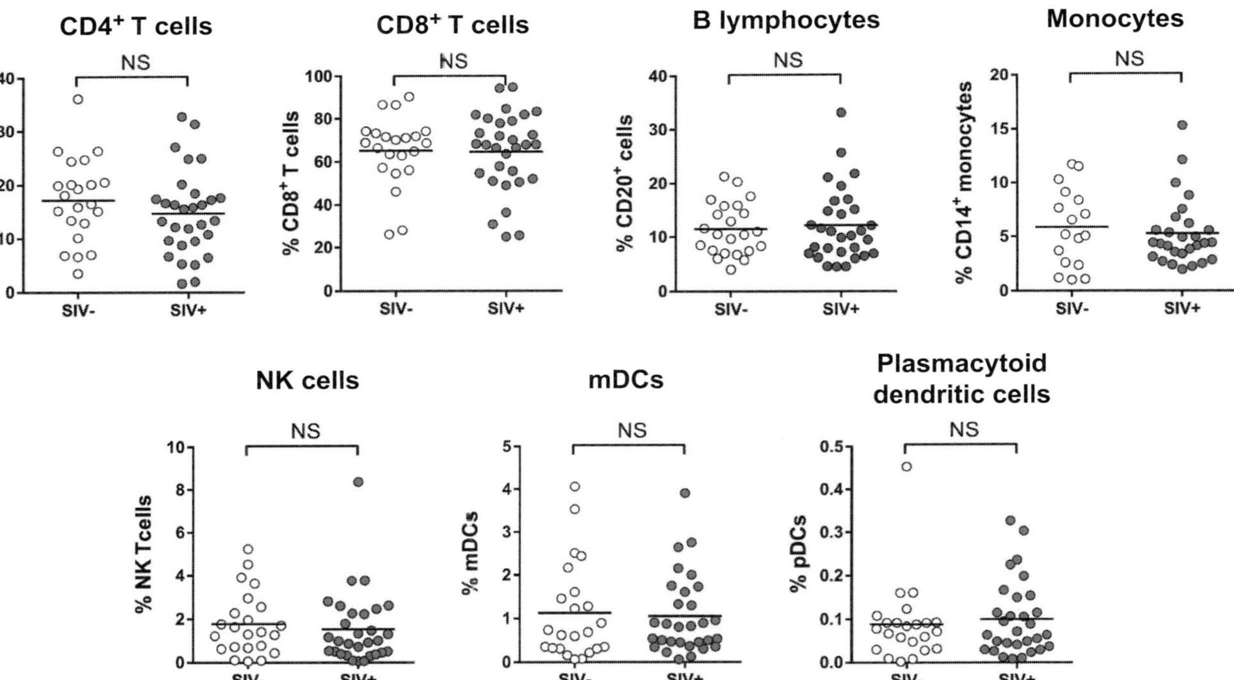

FIGURE 5.2 Cross-sectional analysis of the major immune cell populations in wild *sabaeus* monkeys from The Gambia. No significant difference could be observed between SIV-infected and -uninfected AGMs, suggesting that SIV infection has a minimal impact on immune system homeostasis in AGMs. P-values were calculated by the Mann–Whitney U test. The values on the *y*-axes depict the proportion of the given immune cell population or subset. mDC = myeloid dendritic cell; NK = natural killer; NS = nonsignificant.

from the same cohort of Gambian *sabaeus* monkeys. No difference was observed between SIVsab-infected and -uninfected AGMs (Ma, Jasinska et al., 2014). Furthermore, cytokine and chemokine testing as markers of chronic immune activation in vervet and *sabaeus* AGMs from South Africa and The Gambia showed similar levels of proinflammatory cytokines and chemokines in SIV-infected and -uninfected AGMs, again supporting the nonpathogenic nature of SIVagm infection in the wild (Ma, Jasinska et al., 2014).

CONCLUSIONS

AGMs in sub-Saharan Africa are infected with SIVs at high levels of prevalence. Each AGM species is infected with a species-specific SIV, suggesting host-dependent evolution. SIVs have been present in AGMs for a long time, most likely since the time of their speciation. As a result, SIV infection of AGMs in the wild is largely nonpathogenic. Virus transmission mostly occurs in adult monkeys, primarily through the sexual route. Other transmission routes, such as fighting or grooming, are also involved in SIV transmission in the wild. MTIT is a route of transmission, but these rates of transmission are very low.

6   The Vervet Microbiome

Katherine R. Amato

The rich body of literature focused on vervet biology makes vervets an excellent model for microbiome research. Not only does our knowledge of vervet biology provide important context for understanding how factors such as host diet, physiology, and behavior relate to the gut microbiota, but eventually, it will allow us to more directly examine the mechanisms via which host and microbe interact. However, the field of vervet microbiome research is still in its infancy and much remains to be learned before we can address questions of mechanism. Currently, the published literature describes the microbial communities of two vervet body sites: the gut and the vaginal tract. Here, I summarize what we know about the vervet gut and vaginal microbiota and suggest some future directions.

THE GUT MICROBIOTA

The gut microbiota is the most widely studied host-associated microbial community. A range of studies has reported its influence on host nutrition, health, and diet, and associations between characteristics of the gut microbiota and human disease are rapidly accumulating. For example, data from humans and mice implicate the gut microbiota in the development of host obesity and metabolic disease (Larsen et al., 2010; Ley, 2010; Hosseini et al., 2011). Paired with reduced levels of activity, a high-fat, low-fiber diet is known to increase the risk of human obesity and metabolic disease (Hu et al., 2001), and this diet is consistently associated with altered microbial communities that contain higher relative abundances of some of the taxa known to contribute to obesity and metabolic disease (Yatsunenko et al., 2012; Ou et al., 2013; Schnorr et al., 2014; Clemente et al., 2015; Obregon-Tito et al., 2015). Therefore, researchers have turned toward animal models to test the mechanisms driving host–gut microbe interactions in the context of obesity and metabolic disease. While the use of rodent models is particularly popular due to ease of manipulation, nonhuman primate models are being integrated as well, as a result of their close phylogenetic relationships with humans and presumably similar physiology (Kisidayova et al., 2009; Turnbaugh et al., 2009; Faith et al., 2011; Ma, Jasinska et al., 2014).

In this context, a recent study produced preliminary data describing the effect of a high-fat, low-fiber diet on the vervet gut microbiota (Amato et al., 2015). Sequencing of the 16S rRNA gene was used to describe the gut microbiota of 15 wild vervet monkeys (St. Kitts) consuming ripe fruits, flowers, and seeds (Chapman et al., 1988) and 23 captive vervets (Wake Forest University Primate Center) consuming a high-fat, low-fiber diet (LabDiet 5L0P, Purina, St. Louis, MO; 18 percent protein, 37 percent fat, 45 percent carbohydrates, 9 percent fiber) for six months before sampling. Results indicated that both vervet groups had similar gut microbial richness, but gut microbial community composition differed (unweighted UniFrac distance: $F_{1,26}$ = 4.8, p < 0.001, r^2 = 0.16; weighted UniFrac distance: $F_{1,26}$ = 9.6, p < 0.001, r^2 = 0.28). Specifically, vervets consuming a high-fat, low-fiber diet had significantly lower relative abundances of microbial phyla such as Firmicutes, Lentisphaerae, Proteobacteria, Tenericutes, and Verrucomicrobia and significantly higher abundances of Bacteroidetes and TM7. At the genus level, vervets consuming a high-fat, low-fiber diet possessed lower relative abundances of *Clostridium* and higher relative abundances of *Desulfovibrio, Prevotella, Catenibacterium,* and *Collinsella.*

Because the effect of diet on the human gut microbiota is so strong (Lozupone et al., 2012), these results are not surprising. However, when these data were directly compared to analogous data from humans (Yatsunenko et al., 2012; Schnorr et al., 2014), some interesting patterns emerged. While some members of the gut microbial communities of both vervets and humans responded similarly to differences in host diet, key differences existed (Amato et al., 2015). While microbial richness was unaffected by diet in vervets, in humans, a high-fat, low-fiber diet was associated with reduced microbial richness. Also, while the relative abundances of some of the same microbial taxa differed across diet groups in both vervets and humans, they differed in opposite directions. While people consuming a high-fat, low-fiber diet exhibited increased relative abundances of Firmicutes and reduced relative abundances of Bacteroidetes compared to people consuming a low-fat, high-fiber diet, vervets exhibited a reduction in Firmicutes relative abundances and an increase in Bacteroidetes relative abundances on a high-fat, low-fiber diet. Furthermore, high-fat, low-fiber diet human populations exhibited lower relative abundances of *Prevotella* and higher relative abundances of *Bacteroides*, while the opposite was observed in vervets. As a result, the vervet gut microbiota was more similar to that of humans consuming a low-fat, high-fiber diet, regardless of diet treatment.

Potential confounding factors should be explored in additional studies, and data for other nonhuman primate taxa must be collected, but these results suggest that the human gut microbiota and its response to a high-fat, low-fiber diet differ fundamentally from the nonhuman primate gut microbiota. Two obvious explanations exist. First, it is possible that the unique properties observed in the human gut microbiota are simply the result of unique human diet and physiology. Human evolution appears to reflect a major dietary shift from a predominantly plant-based diet to an increasingly carnivorous diet, with relatively recent increases in food digestibility due to tool use, cooking, agriculture, and other processing techniques (Teaford & Ungar, 2000; Leonard et al., 2007; Carmody & Wrangham, 2009). Additionally, compared to nonhuman primates, humans are characterized by a number of unique physiological adaptations, including increased brain size, reduced gut size, increased fat deposition, and decreased muscle mass (Aiello & Wheeler, 1995; Kuzawa, 1998; Leonard et al., 2003). Since both host diet and physiology drive gut microbiota composition (Dethlefsen et al., 2006), evolutionary changes in human diet and physiology could have led to a distinct gut microbiota.

It is also possible that, in addition to other factors such as diet, unique properties of the human gut microbiota contributed to the evolution of specialized human physiology. If the human gut microbiota conferred an increased capacity for energy production and storage, it could have promoted increased brain size during human evolution. Similarly, because the human gut microbiota plays a role in regulating host energy intake and fat production (Backhed et al., 2004) and an increased capacity to store energy as fat has been hypothesized to have enabled humans to develop larger brains (Kuzawa, 1998; Leonard et al., 2003), changes in the gut microbiota that affected host metabolic pathways could have contributed to the evolution of the human brain as well.

To distinguish between these two alternatives, further studies are necessary that measure the metabolic potential of the human and nonhuman gut microbiota and more directly compare the physiological consequences of consuming a high-fat, low-fiber diet. However, in either case, since microbes like Firmicutes are reported to alter host metabolism and increase fat deposition (Turnbaugh et al., 2006), the microbial patterns observed provide indirect evidence that the human gut microbiota may increase susceptibility to obesity and metabolic disorders in conjunction with a high-protein, high-fat diet. This information could be invaluable to microbiome-focused biomedical research and underscores the need for additional gut microbiome research in vervets and other nonhuman primates.

THE VAGINAL MICROBIOTA

Compared to the gut microbiota, the vaginal microbiota is less commonly featured in the published literature. However, a better understanding of the vaginal microbiota is essential for integrating gut microbes into our theories of host evolution. Not only are vaginal microbes important to host reproductive health (Ma et al., 2012) and therefore fitness, but also they are likely to facilitate the coevolution of hosts and microbial communities at other body sites. For microbes associated with any body site to be implicated in host evolution, they must possess an element of heritability. Because mammals are inoculated by maternal microbes during birth (Dominguez-Bello et al., 2010), the vaginal microbiota is likely to play a key role in the transmission – or heritability – of microbial communities.

While progress has been made in understanding the host-associated factors influencing the composition of the human vaginal microbiota among individuals and within individuals across time (Ravel et al., 2011; Romero et al., 2014), there is a dearth of information regarding the nonhuman primate vaginal microbiota. Nevertheless, a meta-analysis of the vaginal microbiota of several nonhuman primate species and humans was recently completed to evaluate evolutionary trends (Yildirim et al., 2014). Results indicated that humans have a unique vaginal microbiota that is low in microbial richness and dominated by *Lactobacillus*. Additionally, among primates, including humans, vaginal microbiota composition was host-specific but not correlated with host phylogeny. Instead, factors such as geographic origin, female swelling and promiscuity, social structure, testes, body and neonate weight, gestation duration, baculum length, and group size were found to be determining factors.

Samples from six wild vervets (St. Kitts) and six captive vervets (Wake Forest University Primate Center) were included in this meta-analysis (Yildirim et al., 2014). Therefore, we can make some general comparisons between the vervet vaginal microbiota and those of other nonhuman primates and humans. First, similar to other non-human primate species and unlike humans, the vervet vaginal microbiota was not dominated by *Lactobacillus*. In its place, *Sneathia* was identified as a characteristic microbe of the vervet vaginal tract. This was true for both captive and wild vervets, and captivity had no effect on the composition of the vervet vaginal microbiota (pseudo-t = 0.94, p = 0.50). *Sneathia* was also identified as a characteristic microbe of the olive baboon and chimpanzee vaginal microbiota. Together, these data suggest that while *Sneathia* may be associated with vaginal dysbiosis and disease in humans (Harwich et al., 2012), it is likely to be a normal member of the nonhuman primate vaginal community.

Interestingly, in contrast to other nonhuman primates, vervets had low vaginal microbial richness (Yildirim et al., 2014). Although vervets still had a more diverse vaginal gut microbiota than humans (220 operational taxonomic units [OTUs] compared to 67 OTUs), their relatively low vaginal microbial diversity was more similar to humans than other nonhuman primates. Likewise, the composition of the vervet vaginal microbiota was more similar to that of humans and chimpanzees than other Old World monkeys. Additional studies are necessary to determine which common aspects of the host ecology might contribute to these similarities. A unique human vaginal microbiota is hypothesized to be related to factors such as frequency of ovulation, frequency of receptivity, physiological variation, and drift in microbial communities (Stumpf et al., 2013). Evaluations of human and vervet vaginal microbiota in more individuals occupying a wider range of contexts may shed light on the importance of these and other factors in determining the composition of the vaginal microbiota.

CONCLUSIONS

Despite a paucity of microbiome data in the vervet literature, interesting trends have already been uncovered, especially in comparison to the human microbiome. When compared to nonhuman primates, human microbial communities appear to possess unique characteristics in both the gut and the vaginal tract. While these findings may have important implications for our understanding of human health, ecology, and evolution, they are also likely to alter perspectives on nonhuman primates.

For vervets in particular, physiological similarities to humans have made them an attractive model for biomedical trials (Jasinska et al., 2013). However, the results presented here indicate that vervets may not be appropriate models for directly testing the mechanisms behind human–microbe interactions. Even for the vaginal microbiota, vervets appear to represent the best from among a number of flawed models instead of a truly good model. Nevertheless, vervets and other nonhuman primates are still likely to be a critical part of microbiome-centered biomedical research. Instead of replicating the processes occurring in humans, vervet microbiome research can be used to determine whether humans are more susceptible to certain illnesses compared to other primates as a result of their unique micro-biota. If they are, then using models such as vervets to pinpoint the characteristics of the microbiome that are protective will be essential to translating findings into health applications in a clinical context.

In this sense, I believe that some of the most exciting future directions for vervet microbiome research involve biomedical research and the use of vervets as imperfect models for human–microbe interactions in the context of disease. However, studies examining the effects of microbes on wild vervet nutrition, health, and behavior are equally important. Not only will this knowledge enrich existing theories of vervet ecology and evolution, but it will also provide important comparative data for other primates and broaden our understanding of primate ecology and evolution more generally. While microbiome researchers commonly assume that microbes coevolved with their hosts, few studies actively measure host fitness outcomes in response to microbial variation. The rich network of researchers studying vervets in a variety of contexts provides a powerful tool for collecting these data.

Part III **Population Genetics**

7 Population Genetics and Savanna Monkeys

Trudy R. Turner, Christopher A. Schmitt, and Jennifer Danzy Cramer

JG, a large male monkey, was missing. No one had seen him for nearly two months. Males migrate between groups; often they go to neighboring groups. Along the five-mile stretch of the Awash River in central Ethiopia, there was no sign of him. One day, he reappeared. Where had he been? What had he been doing? Why was he back? How long would he stay? Would he become the dominant male in the group? JG's adventures epitomized what we were trying to understand by examining the genetics of these animals: What was the reality of migration and gene flow? JG and his movements gave concrete meaning to the forces of evolution in maintaining these groups as a single genetic entity.

A BRIEF HISTORY OF PRIMATE POPULATION GENETICS

While taxonomy provides the scaffolding for understanding the relationships between organisms, evolutionary theory – and its quantitative ancillary, population genetics – provides the mechanism for understanding the processes by which organisms change over time. Since the initial recognition that allelic differences occur at genetic loci, geneticists have attempted to determine the extent of genetic variation in individuals and in populations. The history of primate population genetics began as soon as the zymogram technique (using electrophoresis and histochemical staining) was introduced. The earliest work on primates was based on an examination of blood samples from animals in captivity. Researchers were asking the very basic question: Do primates have the same types of variation found in humans? This took on a number of very focused studies asking limited questions: Are hemoglobin variants found in baboons (Barnicot et al., 1967; Weiner & Moor-Jankowski, 1969; McDermid et al., 1973)? Or are transferrin variants found in macaques (Goodman et al., 1965)? Often only very few captive animals were sampled. Once such genetic variation was established at protein loci, population genetics and systematic studies became possible. Large-scale surveys of various species from multiple locations were conducted, but provenance could only be estimated roughly as samples were typically obtained from local trappers who captured animals for export. For example, in some of the early work by Darga and colleagues (1975) and Weiss and Goodman (1971), among others, animals were said to come from the general – and largely uninformative – region of Kuala Lampur, Malaysia. Nevertheless, these studies, taken on the whole, demonstrated that there were detectable genetic differences in the frequencies of alleles at various loci not just among species, but also among populations within widely distributed species. Additionally, rates of heterozygosity appeared to approach levels found in humans, although these rates differed in different species. Studies of macaques and baboons in particular, because of their wide distribution, revealed differences that began to elucidate larger evolutionary processes. For example, differences were found between macaque species living on islands and those on the mainland, aiding in the process of understanding genetic bottlenecks and genetic drift. There were many studies that attempted to obtain estimates of genetic differentiation between populations, species, and genera (e.g., Kawamoto et al., 1982). At the time, however, genetic studies of nonhuman primates designed to elucidate microevolutionary processes faced substantial difficulties. As Duggelby stated in 1978:

> Accurate demographic data are difficult to obtain, behavioral data
> may be sparse or conflicting, sampling errors may bias results
> and an unknown amount of gene flow may take place from
> contiguous populations ...

This stage of primate genetic research was quickly followed by a series of studies conducted to explicitly address the concerns articulated by Duggelby. In this new wave of studies, animals of known groups were trapped, precise provenance was known, and demographic data were collected. As an outcome of these more precise demands, researchers were necessarily responsible for collecting all primary data themselves and no longer relied on commercial trapping. These studies were often targeted to species where there was already substantial behavioral data available (which is the most time-consuming aspect of the new data demands). At the time, populations of several species of Old World monkeys had already been the subjects of long-term behavioral observation. The majority of these taxa were semiterrestrial and so relatively easy to trap. Models of human evolution were in place as well that regarded Old World monkeys to be particularly reasonable analogues of early hominin activity. All of these factors predisposed researchers to focus trapping on baboons (Jolly & Brett, 1973; Olivier et al., 1974; Ober et al., 1978, 1980; Byles & Sanders, 1981; Olivier et al., 1986; Rogers & Kidd, 1993), macaques (Nozawa et al., 1982), and vervets (Turner, 1981; Dracopoli et al., 1983; Shimada & Shotake, 1997) for this next stage of primate genetics studies. Through the 1990s, most studies on Old World primate genetics continued to concentrate on these three well-sampled genera. During that time period, nearly 11,000 macaques, 3000 baboons, and 750 guenons were surveyed (Turner & Weiss, 1999). Within the guenons, savanna monkeys were sampled almost exclusively, while within the macaques, four species – *fascicularis*, *fuscata*, *mulatta*, and *nemestrina* – were sampled extensively. A greater percentage of the macaque data lacked demographic information than did the data for baboons and savanna monkeys, yet macaques were sampled over a greater extent of their range than baboons and savanna monkeys.

One group of these studies focusing on large-scale demography in well-studied taxa was useful in understanding how genetic variation was partitioned between troops, between trapping locations, and between populations separated by greater geographic space (e.g., Dracopoli, 1983; Olivier et al., 1986) and in beginning to understand levels of gene flow between populations. *Macaca fuscata*, for example, are distributed across Japan, but behavioral research suggested that local breeding units were composed of a number of troops connected by frequent male migration. Extensive genetic sampling provided evidence that, indeed, high rates of male migration could be found to have an impact upon allele frequencies across hundreds of troops of *Macaca fuscata* in Japan (Nozawa et al., 1982), but that this variation was geographically constrained, as troops separated by greater geographic distances were genetically more distinct (Nozawa et al., 1982; Williams-Blangero, 1991), although whether this was caused by a lack of gene flow or random genetic drift was unclear. Using a different set of techniques for estimating variability – immunoglobin allotypes – Coppenhaver and Olivier (1986), on the other hand, found that the highest variability among baboons at widely dispersed locations in Kenya was between local populations. This same pattern of intergroup differentiation was also seen in baboon troops sampled by Rogers at Mikumi Park in Tanzania (Rogers, 1989). Differences found between groups were attributed to the sampling of the small number of male migrants that do not necessarily represent local gene pools (Williams-Blangero, 1991; Rogers, 2000). The pattern of genetic variability among geladas (*Theropithecus gelada*) was different from that of the closely related *Papio*. Four groups living in the Simien Mountains of Ethiopia were sampled and were found to be genetically homogeneous. Shotake and Nozawa (1984) postulated that a bottleneck that limited genetic variation in the original founding population could explain this lack of differentiation.

Even though population structure and demographic data were available for the animals in these studies, there was still limited accompanying behavioral information. The next group of large-scale studies were designed to resolve this disconnect (Melnick & Kidd, 1983 and Melnick et al., 1984 on *Macaca mulatta*; de Ruiter et al., 1992 and de Jong et al., 1994 on *Macaca fascicularis*; and Phillips-Conroy et al., 1992 on *Papio hamadryas*). Like other large-scale surveys, these studies used Wright's fixation index, or F_{ST} (the observed variance in subpopulations relative to the maximum theoretical variance of completely isolated subpopulations), to measure the relative degree of differentiation among subgroups. They found population differentiation to be low overall in these groups. This relatively low level of differentiation was regarded as the result of high rates of gene flow between groups mediated by observed male dispersal events (Melnick & Pearl, 1987). The additional behavioral information also allowed for added depth to the genetic analysis. Melnick and colleagues, for example, found little to no genetic variation among daughter groups that had fissioned from a single ancestral group in *Macaca mulatta* (Melnick & Kidd, 1983). De Ruiter and Geffen (1998) were, in turn, able to demonstrate that, in *Macaca fascicularis*, high-ranking males mated more frequently with high-ranking females in a single social group.

The popularization of microsatellite analysis caused an explosion of population genetics research in primates during the late 1990s and 2000s, yielding a wide array of studies in an unprecedented number of primate taxa, including chimpanzees (Gagneux et al., 2001), orangutans (Kanthaswamy & Smith, 2002), snub-nosed monkeys (Li et al., 2003), howler monkeys (Cortes-Ortiz et al., 2003), and sifakas (Lawler et al., 2003), to name just a few. However, more recent microsatellite studies on primate genetics have concentrated less on assessing local fluctuations of gene frequencies and more on attempting to answer refined genetic questions associated with behaviors, such as paternity in a variety of organisms (Martin et al., 1992), including mouse lemurs (Andres et al., 2003), baboons (Altmann et al., 1996; Buchan et al., 2003), rhesus macaques (Keane et al., 1997; Widdig et al., 2001, 2004; Bercovitch et al., 2003), Barbary macaques (Menard et al., 2001), langurs (Launhardt et al., 1998), chimpanzees (Vigilant et al., 2001), bonobos (Gerloff et al., 1999), and even vervets (Newman et al., 2002); male care and incest avoidance (Erhart et al., 1997; Alberts et al., 1999; Menard et al., 2001; Buchan et al., 2003); the relationship between kin and social interactions (Goldberg & Wrangham, 1997; Alberts, 1999; Mitani et al., 2000; Widdig et al., 2001; Lukas et al., 2005; Langergraber et al., 2007);

and how molecular data can be used to address the often difficult to observe behavior of dispersal (Di Fiore, 2009; Di Fiore et al., 2009).

One of the best-characterized populations of Old World monkeys, both genetically and phenotypically, is that of the macaques found on Cayo Santiago, off the coast of Puerto Rico. From a founding population of 409 macaques from India, brought to the island in 1938, a thriving, semi-free-ranging population has flourished more or less independently on the island. The population has been under close monitoring since 1956 and has become a powerful data set for genetic analysis over the past 40 years (see Widdig et al., 2016a for a comprehensive review). Periodic censuses have documented the history of the colony with several ecologically mediated declines and expansions. Since the colony has been closed to migration, all potential mates are known and pedigreed (Williams-Blangero, 1991), providing a model for understanding population processes and evolution. Indeed, the macaques on Cayo Santiago have been used to gain a greater understanding of a wide array of systems, from socially mediated population processes like male reproductive skew (Berard et al., 1993; Dubuc et al., 2011) and paternal kin bias (Widdig et al., 2016b), to the quantitative genetics underlying life history trade-offs such as age at first reproduction and age of death (Blomquist, 2009), to more biomedical questions like the functional genomics of adult-onset macular degeneration (Pahl et al., 2013).

Some wild populations of Old World monkeys have also now been studied for decades, and a rich history of behavioral information is available. Perhaps the best known are the baboons (*Papio cynocepahalus*; and the neighboring *Papio anubis* and their hybrids) that have been studied by the Amboseli Baboon Research Project (https://amboselibaboons.nd.edu) in Amboseli National Park, Kenya. Jeanne and Stuart Altmann began studying these baboons in 1963 and have been consistently working with the population since 1971. Although the majority of these data focus on intensively sampled, noninvasive behavioral research, beginning in 1989, the group began trapping in order to collect blood samples for genetic and hormone data, and pioneered much early work linking genetic variation to behavioral processes (e.g., Altmann et al., 1996; Bayes et al., 2000). More recently, the work of Jenny Tung and Marie Charpentier has carried the Amboseli Baboon Research Project from microsatellite-based population studies (e.g., Charpentier et al., 2012; Tung et al., 2008) to novel and exciting next-generation genomic methods applicable to field behavioral studies in primates (Tung et al., 2010), including analyses of gene expression (Tung et al., 2011) and hybridization with *Papio anubis* (Charpentier et al., 2012).

The revolution in genetic and genomic technology over the past decade has tantalized primatologists. In many ways, it has provided answers to many theoretical questions – however, the difficulties in obtaining samples, even noninvasive samples, has largely limited the total scope of what could be accomplished. This is especially true with regard to whole-genome sequence-based analyses that require higher-quality genomic data than microsatellite analyses (although work is still ongoing to extract next-generation sequencing [NGS]-quality genomic data from noninvasively collected feces; e.g., see Perry et al., 2010). Despite these difficulties, primatological field studies have consistently managed to incorporate NGS genomic technologies and analyses into their studies, and such work has already become a large presence in field primatology (e.g., Tung et al., 2010; Bradley & Lawler, 2011; Kelaita, 2015). Savanna monkeys are no exception. The International Vervet Research Consortium (IVRC) has incorporated genomics into our research, including research in gene expression (see Jasinska, this volume) and viral–host phylogenomics (see Apetrei and colleagues, this volume), in addition to having the most contiguous reference genome of any nonhuman primate (Warren et al., 2015), enabling unprecedented confidence in analyzing structural variation in the genome and what influence it may have on population-specific phenotypes (Jasinska, this volume; Warren and Montague, this volume).

SAVANNA MONKEY GENETICS

The history of research on genetics in savanna monkeys is similar to that of other cercopiths. Animals housed in captivity or of roughly known provenance were first sampled to establish heterozygosity at protein loci (Ishimoto et al., 1967; Coppenhaver & Buettner-Janusch, 1970; McCombs & Bowman, 1970; Barnicot & Hewett-Emmett, 1971, 1972; McDermid & Ananthakrishnan, 1972; Downing et al., 1973; McDermid et al., 1973). The most extensive sampling, of 61 animals, was conducted on laboratory animals in South Africa. Provenance of these animals was reported as either Transvaal/ Bulge River or Natal – both major provincial regions of the country. Heterozygosities occurred only at a few loci and levels were always low, suggesting a general lack of genetic diversity in the populations sampled.

Even though population genetic studies were underway for baboons and macaques, savanna monkeys were unstudied in the wild until Turner and Jolly began work in Ethiopia in the early 1970s (Jolly et al., 1977; Turner, 1981). Since that time, there have been an additional four population genetics studies of savanna monkeys – three of which were conducted by Turner in collaboration with her students and colleagues. The first study, at Awash Park in Ethiopia, was conducted at the same site as the well-known Awash baboon project. Collections of samples were, in fact, conducted at the same time using similar methodologies. The second study, conducted by Turner, Dracopoli, and Jolly, was initiated in Kenya several years later as an extension of the Ethiopian study and was designed to sample a region more central to the animal's known range. It included animals at four major sites and became the model for the third study in South Africa, in that sampling took place across the animal's range, but also occurred intensively at each major site. The final study was part of the International Vervet Genome Project, in which animals were sampled from throughout their range in Africa and the Caribbean. Understanding the processes of evolution in savanna monkeys – migration, gene flow, genetic drift, and selection – was a key aspect of all of these projects. Unlike largely phylogenetic population projects that look to validate the taxonomic products of evolution, these projects tried to understand the processes of evolution as they occurred within these particular populations and the larger divisions within the genus *Chlorocebus*.

In all, these projects took place over a 40-year time span. Over that time, not only has access to genetic information changed, but so have statistical techniques and the evolutionary questions that these techniques seek to address. Here, we present a brief overview of these projects, including an additional project on the population genetics of vervets in Ethiopia.

VERVET GENETICS: ETHIOPIA

In the initial Turner study, 124 animals in seven social groups living along the Awash River were captured and sampled.[1] This number represented 93 percent of the animals known to live in Awash National Park at that time. In addition to blood samples, saliva samples, weight, and body length data were collected, and age and sexual maturity were determined. Of the 23 loci examined by starch gel electrophoresis, 13 were invariant and 10 showed some level of variation. In addition, the ABO-like blood group locus exhibited variation. Only 17 percent of the loci were polymorphic and the average heterozygosity was 5.6 percent, a level consistent with the average heterozygosity found for protein loci in other vertebrate species examined at the time (Selander & Kaufman, 1973). Using Wright's F_{ST}, Turner found that the average for all loci was 0.062, a relatively low value, which was confirmed by a chi-square test for group homogeneity. In other words, the seven groups sampled were not genetically

---

[1] An additional 125 vervets housed at laboratories and breeding facilities in the USA were also sampled. Some pedigree data were available for these colony animals. It was therefore possible to determine the inheritance pattern of some variant phenotypes found in these colonies.

differentiated. Wright's Island Model indicated that the amount of migration between groups was one to two individuals per year. This corresponded to behavioral estimates of vervet group transfer and suggested that all seven groups sampled comprised a single breeding population. Using Wright's model for effective population size (N_e), it was estimated that only about half of the animals in a given group were part of the breeding pool. If N_e for a group was on average 8 and if one to two individuals were migrating per year, the rate of migration was about 25 percent of the group per year. This high proportion of migratory individuals explains the lack of differentiation between local groups found along the Awash River.

Vervets in the Awash River Valley have been revisited by others in the years since the initial study by Turner. Shimada and Shotake (1997) collected 196 blood samples from individuals in 11 troops along a 600-km expanse of the Awash River in central Ethiopia. The rate of polymorphic loci (out of a total of 33 loci examined) was 17 percent and the average heterozygosity was 5 percent, a level of variation directly comparable to that found by Turner. Shimada and Shotake included not only variation determined by starch gel electrophoresis, but also variation detected by the use of iso-electric focusing (IEF). F_{ST} for these populations was 0.075, also similar to Turner's value of 0.062, even though the geographic distances between groups were much greater than in Turner's sample. These similar values are, however, lower than those found in comparable populations of *M. fuscata* (Nozawa et al., 1991) and in neighboring troops of *M. fascicularis* (Kawamoto et al., 1984). When this same set of samples was later examined by restriction fragment length polymorphism analysis of the mitochondrial genome, Shimada (2000) identified ten haplotypes in five clusters or haplogroups. These haplogroups were distributed in blocks that ranged between 120 and 250 km along the Awash River. Diversity within haplogroups was not great, while diversity between groups was more significant. Haplogroup blocks were more clearly defined for females, although blocks in highland areas were not as clearly defined as in lowland areas. In lowland areas, migration was restricted to riparian forest along the Awash River, as opposed to the more widely forested highlands. The distribution of male haplogroups suggested that males migrated between the blocks. When sequence diversity in the mitochondrial control region was examined, Shimada and colleagues (2002) found that divergence within what they called a single subspecies in their sample was comparable to that observed between subspecies of vervets.

Shimada and Shotake (1997) suggested – based on paleoclimatic and paleoecological data – that the area around the Awash River has been subjected to fluctuations of water level, temperature, and humidity. Periods of cold and dry climate may have historically reduced the numbers of animals in the area, while warm and wet climatic conditions may have led to occasional population expansions. They further suggested that vervets recolonizing an area during these warmer and wetter periods may have adopted an r-selected strategy of rapidly reproducing and thus increasing population numbers, which would have effectively pushed vervet populations through a series of bottlenecks followed by rapid expansion. The overall effect of this process would be that, even though populations might reach large numbers, the genetically detectable effective population size of sampled groups would remain low. In other words, this series of environmentally mediated bottlenecks would continually reduce genetic variability, and would happen with enough frequency to prevent the recovery of genomic diversity by mutation or drift.

Would these low levels of variability occur in vervets in other parts of their range? The vervets of central Ethiopia were at an extreme end of the species range in East Africa. Further east and south, the environment turned more desert-like, conditions in which it would be more difficult for vervets to survive. What would happen if we examined vervets that lived more centrally in the taxa's range? Would we see the same levels of population variation and population differentiation? Would genetic structure follow geographic ranges in the same way as in Ethiopia?

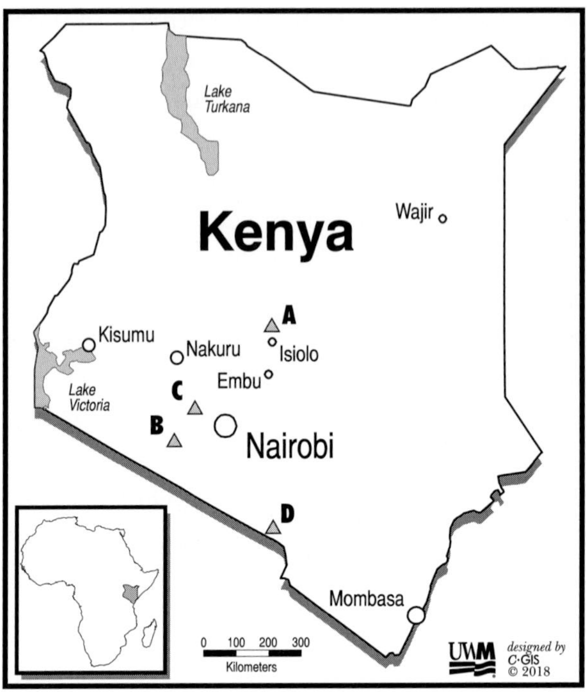

FIGURE 7.1 Sampling locations in Kenya, 1978–1979. Each location represents between five and nine troops and between 50 and 125 animals. A = Samburu; B = Mosiro; C = Naivasha; D = Kimana.

VERVET GENETICS: KENYA

The Kenya study was designed to address these questions by comparing genetic variation at three levels: within a given social group, among social groups at sites where animals could move easily between local social groups, and between sites separated by significant geographic distances (see Figure 7.1). In addition to the measures collected in Ethiopia, a much broader series of morphometric measurements and tooth casts were collected from over 360 animals living in over 30 troops at four locations – Samburu, Naivasha, Kimana, and Mosiro – that were separated by between 80 and 300 km.

Initially, in accordance with the methods available at the time and consistent with the initial study in Ethiopia by Turner, serum proteins and ABO-like blood groups were examined (Dracopoli et al., 1983). Three of nine serum proteins (transferrin, group-specific component, and prealbumin carboxylesterase) examined by protein electrophoresis were polymorphic, as was the ABO-like blood group locus. The proportion of polymorphic loci (17 percent) and average heterozygosity (0.05) were similar to those found in Ethiopia.

The Kenya study was able to take advantage of statistical techniques that partitioned genetic diversity into several basic components – within troop, among troops in a geographically distinct area, and between these distinct areas or sites. The technique was the G_{st} statistic of Nei (1972, 1973) and Chakraborty (1974), which is an extension of Wright's F_{st} and other F statistics. Allelic diversity within each site (H_s) is divided into allelic diversity within individual troops (H_c) and among troops at the same site (D_{cs}). Overall allelic diversity (H_t) and allelic diversity among troops at different sites (D_{st}) are calculated from weighted means of allele frequencies at each site (Dracopoli et al., 1983). Total allelic diversity (H_t) is equal to the sum of these three components ($H_c + D_{cs} + D_{st}$), from which it is possible to apportion genetic variability. Out of a total of 100 percent, within-troop variability was calculated at 83 percent, between troops at the same site at 8 percent, and between different sites at 9 percent. This same level of homogeneity between sites was confirmed by F_{st} and G_{st} (Dracopoli et al., 1983).

Euclidian distance measures were also applied to troops at the various sites to test whether allelic diversity corresponded to geographic distance using a two-dimensional MDSCAL diagram. MDSCAL is a non-metric, multidimensional scaling technique for the analysis of similarity and dissimilarity and finds the best geometric representation for data in space. Using this technique, Mosiro and Kimana – two sites in the south – are each tightly clustered, while Samburu and Naivasha in the north are more widely distributed and actually overlap each other. Naivasha has the most geographically dispersed troops – the territories of these troops run along the edge of Lake Naivasha, the shores of which spread vervet troops about

five times further apart than the distances between troops at any of
the other sites. The vervets around the Naivasha shore were further
distanced by occupying patches of forest between agricultural lands.
In Samburu, even though the troops lived along a short course of the
Samburu River, there was one troop that had crossed a bridge to the
other side, ranged and fed on the opposite bank, and thus expanded
their distance from the other troops. This was indicated graphically.
Notably, the river-crossing troop in Samburu had the opportunity to
mate with animals on the other side of the river; this is an opportunity for novel allelic variation that may be absent in the other
troops.

The relatively low levels of inter-troop and inter-site differentiation in Kenya require some explanation. We could postulate – as
Shimada does for Ethiopian populations – that these groups also under-
went ecologically mediated rapid expansions from small founding
populations. Climatic conditions in Ethiopia, however, may be more
amenable to this explanation than those in Kenya. Conditions in
Ethiopia were markedly and consistently drier than those in Kenya.
The relative homogeneity of the populations in Kenya could be due
to high levels of gene flow, but such an explanation requires either
a continuous distribution of vervets or potentially quite large migration distances. In Kenya, two sites are highland sites, another two are
lowland sites. Urban areas occur between sites. Cheney and Seyfarth
(1990a) found that vervets in Amboseli Park, Kenya, usually do not
move far from their natal groups, and males often go to neighboring
groups where their brothers have migrated. Although possible, these
explanations seem equally unlikely. Selection may also provide an
explanation: the particular proteins examined in this analysis may be
conserved by natural selection, resulting in the observed homogeneity. One way to resolve these questions is to examine loci known to
be both selectively neutral and highly variable, as opposed to protein
products that may be conserved by selection for function. The advent
of microsatellite and sequencing techniques would have made this
possible, but would have to wait more than a decade to occur.

The original Kenya samples were collected in 1978–1979. It was standard practice at that time to take a sample, centrifuge it to remove the serum, pipette off the buffy coat, and wash the red blood cells three times in an isotonic (physiological) saline solution. The erythrocyte mass was then stored and shipped to labs in the USA and London, UK, for long-term storage and analysis. Anyone familiar with DNA extractions will immediately note that we threw away the white cells, the DNA-rich portion of the blood from the sample. We literally had to wait until the development and popularization of the polymerase chain reaction (PCR) to study additional genetic variation in these samples. PCR can amplify even tiny amounts of sample – a few cells left in the erythrocyte mass. This left us with about 7 μg of DNA for each animal, just enough to examine microsatellite loci. Of the dozen microsatellite loci originally examined, three were highly polymorphic. There are several questions that we asked: What is the level of variability in microsatellite markers? Could sites and troops be differentiated on the basis of these markers? How did microsatellite variation compare to the variation previously observed for electrophoretic proteins? How did the pattern of differentiation observed in microsatellite markers compare to the differentiation seen in morphological features?

In order to explore diversity within troops and sites using microsatellite markers, a replicated test of goodness of fit was used (Sokal & Rolf, 1981). First, we calculated the G statistic for heterogeneity (G_h) for the 21 troops where microsatellite data were available for more than 50 percent of the animals, effectively reducing the total sample size to 240 animals. The second step was to generate an overall distribution of alleles. This distribution was used to assess the lack of fit between individual troops and the expected distribution derived from the pooled data (G_i). Troops that differed greatly from the pooled data distribution were examined to determine which alleles were over- and under-represented. There were considerable differences in the number of alleles generated at each of the polymorphic loci; the greater the number of alleles at a locus, the greater the likelihood that sites and troops would be heterogeneous. On the other hand, under conditions of high polymorphism and small group size, low-frequency alleles would not be present in some groups. In addition, rare alleles might have a disproportionate effect on distance statistics. Given this, it is possible that the absence of an allele in any particular troop or location did not reflect true evolutionary processes, but rather was a sampling error.

Several statistical techniques were employed to compare patterns of differentiation generated by electrophoretic techniques with patterns generated by microsatellite loci. These included the ordination technique, constant row total–multiple correspondence analysis (CRT-MCA). This statistic is equivalent to F_{st}, but in this specific case differentiation can be apportioned to particular loci. The technique estimates the degree of population differentiation, considers all loci simultaneously, and determines the role of each locus in the pattern of population differentiation. When looking at the electrophoretic data and the ABO-like blood group data together, Kimana and troop Samburu 5 are strongly differentiated from the other troops, especially those at Naivasha. The ABO-like blood groups alleles are mainly responsible for differentiating the troops; all of the animals at Kimana had A alleles at the ABO-like locus, as did those of Samburu 5. Variation at this one locus causes these troops to cluster more closely together. The F_{st} values for the microsatellite alleles derived from CRT-MCA are much higher than they are for the protein alleles, primarily because of the number of alleles at each locus. How did this coordinate with the geographic location of the troops? Sambura and Kimana are the furthest apart geographically. An additional two techniques – a Procrustes analysis and a quadratic assignment procedure – were employed to compare these data sets. The Procrustes analysis compares the two-dimensional grouping of troops produced by the electrophoretic data with the two-dimensional grouping produced by the microsatellite data. The electrophoretic coordinates were rotated to the target of the microsatellite coordinates. The resulting Procrustes value, 0.88, is quite far from the zero value that indicates a perfect fit between the two solutions. The two systems do not provide an equivalent grouping of troops. One notable feature, however, is that Naivasha troops 1, 8, and 9 are separated from other Naivasha troops.

The low agreement between the groupings produced by electrophoresis and microsatellite data sets was supported by the use of a quadratic assignment procedure, a non-parametric test of goodness of fit that assesses how well the two results match – the better the fit, the higher the value. The observed correlation between the two data sets was 0.047, which is very low.

We were able to create a geographic structure matrix and compared it first to the electrophoretic matrix and second to the microsatellite matrix. The first attempt distinguished between within-site troops and between-site troops. Troops belonging to the same site were assigned a 0 and those belonging to a different site were assigned a 1. The microsatellite solution had a correlation value of 0.128, while the electrophoretic solution had a correlation value of 0.198, implying that both data sets had some level of geographic structuring. When the same analysis was conducted using a more realistic geographic matrix composed of actual distances between sites and between troops at a site, the electrophoretic data exhibited a significant correlation value of 0.306, while the microsatellite correlation value fell to 0.097.

All of the correspondence analyses indicated that Naivasha troops 1, 8, and 9 formed a tight cluster. When we located these troops on the Naivasha map, we found that these troops were somewhat isolated from the other troops at this site. We created a geographic structure matrix that lumped these three troops together and separated them into a fifth distinct site. Both the electrophoretic and the microsatellite matrices correlated significantly with this five-site matrix, with levels of 0.48 and 0.49, respectively. All of this indicates that there is some microgeographic structuring of troops at sites and that geographic distance alone is not enough to pattern genetic structure in these populations.

Over time, we have been able to add additional microsatellite loci to these analyses and to take advantage of newer genetic and statistical techniques. One of the most useful statistical techniques is derived from landscape genetics. Landscape genetics does not assume a troop or population structure. Each individual is treated as an independent data point and calculations of distance are computed using individual data regardless of troop structure. STRUCTURE is a software program used to investigate multi-locus genotype data and assigns individuals to populations (Pritchard Lab). When this technique was applied to an expanded Kenya data set that included an additional eight microsatellite loci, all of the animals were divided into two major groupings along a north/south axis (Grobler et al., 2012). There was clear genetic differentiation between these locations – minimally recognized by electrophoretic information, but also indicated by mitochondrial DNA (Haus et al., 2013; Turner et al., 2016) and single-nucleotide polymorphism information (see the discussion of the work of Svardal et al. below). Is there something about this geographic area that can help in understanding this patterning? This area of East Africa – from Ethiopia through to Kenya and west through to Uganda – is where three savanna monkey taxa coexist: *Ch. aethiops*, *Ch. pygerythrus*, and *Ch. tantalus*. In all probability, the borders of these taxa are fluid – and all of these borders are relatively near to the trapping sites in Kenya. This particular area is one of dynamic population movement conjoined with some differentiation. Interestingly, the animals at the southernmost site, Kimana, are more similar genetically to animals in South Africa than they are to animals to the north in Kenya (Lorenz et al., 2010). Kenya seems to be a hot spot for savanna monkey evolution.

VERVET GENETICS: SOUTH AFRICA

South Africa became the site of the next major sampling location of vervet monkeys. Many vervets were living in sanctuaries due to the heightened tensions between vervets and people brought on by crop raiding, and the need to attend to the welfare of many of these animals spurred this locational shift in focus. South African wildlife authorities, using some older taxonomic work, had determined that three subspecies of vervets lived in South Africa, and they were reluctant to mix taxa in the sanctuaries. But were there really three taxa in South Africa? Some initial work had been done previously on vervet genetics by South African geneticist J. Paul Grobler and colleagues (2002). To address the concerns of the wildlife authorities, Grobler and Turner collaborated to employ Turner's Kenya strategy of broad sampling combined with intensive local sampling on South African vervet populations. Given the low heterozygosity implied by the microsatellite work in Kenya, intensive local sampling was necessary to avoid missing a significant amount of variation. Sampling began in 2002 and continued through to 2009. The results of this effort are presented in Coetzer and colleagues (this volume).

The South Africa project led to the even more extensive sampling project of the IVRC. With the IVRC, vervet genetic research has expanded into the realm of NGS and genomics analyses. Through IVRC trapping efforts, a subset of samples from animals representing all of the major taxa of savanna monkeys were collected and sequenced (Jasinska et al., 2013; Svardal et al., 2017). Through the work of Svardal et al. (2017), the richness of the NGS data allowed for the discovery of previously unknown historical population dynamics, suggesting that the population history of the genus *Chlorocebus* is one of initial admixture between recently diverged populations followed by gradual differentiation via isolation by distance. But that's not the whole story – using admixture to cluster individuals into groups and D statistics from ABBA–BABA tests to assess shared ancestry and gene flow, Svardal et al. (2017) found that even after the initial split, there was extensive gene flow across populations. Ghanaian *Ch. sabaeus* and East African *Ch. p. hilgerti* showed extensive admixture with *Ch. tantalus*, whose range abuts both, and that this shared ancestry represents gene flow (15.2 and 5.0 percent, respectively). Multiple sequential Markovian coalescent methods and the lack of shared haplotypes across taxa, despite clear shared genetic material, suggest that this gene flow was ancient rather than recent.

What Svardal et al. (2017) call the *Ch. cynosuros–hilgerti/ pygerythrus* complex shows a more complicated and recent history of gene flow. Although morphology suggests that *Ch. cynosuros* and *Ch. pygerythrus* are separate species, while Kenyan/Tanzanian *Ch. p. hilgerti* and South African *Ch. p. pygerythrus* are both subspecies subsumed within *Ch. pygerythrus*, the genomic data suggest that the *Ch. cynosuros* populations may actually be genetically closer to *Ch. p. pygerythrus* than either is to *Ch. p. hilgerti*, with abundant admixture found between these three groups. In contrast to simple clustering, admixture analyses suggest that *Ch. cynosuros* populations are, in fact, admixed populations of *Ch. p. pygerythrus* and *Ch. p. hilgerti*, with larger genetic contributions from the latter. In contrast to the ancient admixture and subsequent separation along equatorial Africa seen in *Ch. sabaeus/tantalus/hilgerti*, the southern expansion of savanna monkeys appear to have admixed as recently as 10,000 years ago (between *Ch. cynosuros* and *Ch. p. pygerythrus*), followed by isolation by distance and potentially continued admixture where these populations met (e.g., Botswana in the case of *Ch. cynosuros* and *Ch. p. pygerythrus*).

Further results from this genomic, epigenomic, and metagenomic research – and future directions – are presented in this volume by Jasinska (regarding the overall structure of the IVRC sampling, strategy, and aspects of savanna monkey genomics and gene expression), Warren and Montague (regarding the savanna monkey reference genome and structural differences across taxa), and Amato (regarding metagenomics sampling and research on the vervet microbiome).

8 Population Genetic Structure of Vervet Monkeys in South Africa

Willem G. Coetzer, Joseph G. Lorenz,
Nelson B. Freimer, and J. Paul Grobler

From a conservation and evolutionary genetic perspective, vervet monkeys were a largely neglected group in South Africa until the late 1990s. This changed when conservation authorities in some provinces expressed concern about the mixing of possibly unique genetic units due to artificial translocation of animals, in what was one of the first applications of the evolutionary significant unit (ESU) concept in South Africa. This interest led to more questions on the evolutionary genetics of vervets, including: "What routes were followed during historical migration of vervets into the region?" and "How do current patterns of genetic diversity reflect the influence of barriers to gene flow in the region?"

A number of landscape features could potentially define genetic connectivity and routes of migration into and within South Africa. Firstly, South Africa is divided into low-laying inland and coastal regions and an elevated interior plateau, separated in part by the presence of the Drakensberg mountain range. The low-laying areas are covered with savanna and thicket biomes, forming uninterrupted vervet habitat. Similarly, on the plateau behind the mountain range, there are extensive areas of savanna biome that form a good habitat for the species in the north and northeast. However, toward the central part of the country and in the western to southwestern regions, the savanna biome makes way for grassland and Karoo (semi-desert) biomes. The central grassland is superficially unsuitable as a habitat for vervets, due to the lack of tree cover and low winter temperatures. Nevertheless, vervet monkeys are found in these grasslands, utilizing tree cover along river courses and local thickets (Turner et al., 2016).

Adaptation to colder temperatures is evident from observations of sunbathing (Danzy et al., 2012). In the Karoo biome, vervet distribution is facilitated by river courses offering tree-covered habitats in otherwise barren landscapes.

We postulate that vervet monkey populations in the savanna and thicket areas of South Africa will display the genetic signature of a continuous model of gene flow and migration and be affected by isolation by distance only. In contrast, populations in the grassland and Karoo biomes would most likely show a stepwise model of migration and gene flow along approximately linear routes of migration following river courses and local areas of suitable habitat.

In order to provide empirical data to test our assumptions on patterns of geographical genetic diversity, we launched an extensive project to sample and genotype wild vervet populations from across the natural dispersal range in South Africa. This included both extensive sampling across the range and intensive sampling at specific regional sampling sites to determine finer-grained estimates of genetic variation. Sampling was conducted from 2002 to 2010 and geographical coverage was designed to reflect the influence of potential barriers to gene flow highlighted above.

Animals were trapped and an ear-punch or biopsy was taken from each individual studied (Grobler & Turner, 2010). We used an approach based on trapping in preference to the alternative of non-invasive use of fecal samples in order to construct a comprehensive data set that goes beyond genetics to also include data on morphological, serological, endocrinological, and other biological variables. Throughout collection, we operated according to ethical clearance provided by our academic institutions. We also received research and collection permits from the provincial environmental affairs departments of all of the provinces where collections were done. During fieldwork, we attempted to convey a conservation message to landowners where collections were made.

Genetic analysis was based on variation at the mitochondrial DNA (mtDNA) control or D-loop region, following many other phylogeography studies on nonhuman primates (e.g., *Lagothrix*: Botero et al., 2015; *Hylobates*: Whittaker et al., 2007). A 460-bp fragment of the region, including part of the hyper-variable region I (HVI) and the central conserved region of hyper-variable region II (HVII), was sequenced (Turner et al., 2016).

We identified 26 distinct mtDNA haplotypes in a total of 101 vervet monkeys originating from 15 populations. In phylogenetic analysis, the populations sampled were grouped into three well-defined clusters. The clusters were populated as follows: (1) a significant portion of populations sampled grouped together in a cluster that contained animals from the savanna biome regions sampled in the northern and northeastern inland sites, as well as populations toward the northwest that fall outside the savanna areas, but are found in riverine thickets along the west-flowing Orange River system. This cluster also shows connectivity to savanna and coastal thicket populations along the northern parts of the Indian Ocean coastal belt via routes of migration around the northern edge of the Drakensberg range. (2) Populations from the central grassland areas show close identity with each other and significant divergence from the cluster described above. It is likely that vervet monkeys colonized the central grassland biome due to sporadic migration along corridors of suitable habitat along a largely treeless grassland biome. Options for continued gene flow with other populations to the north would be restricted to occasional movements, leading to genetic drift. (3) The remaining cluster contains vervet monkeys found along the southern part of the coastal belt in the coastal thicket biome. This cluster is most closely connected to the northern Savanna group rather than the remaining coastal populations further along the coast to the northeast (Figure 8.1). This suggests the presence of a barrier to gene flow along the otherwise homogeneous coastal belt, possibly centered on the Tugela River. Bayesian evolutionary analysis used to supplement the classic phylogenetic analysis showed that the three clusters diverged between 0.815 and 1.099 million years ago (Turner et al., 2016).

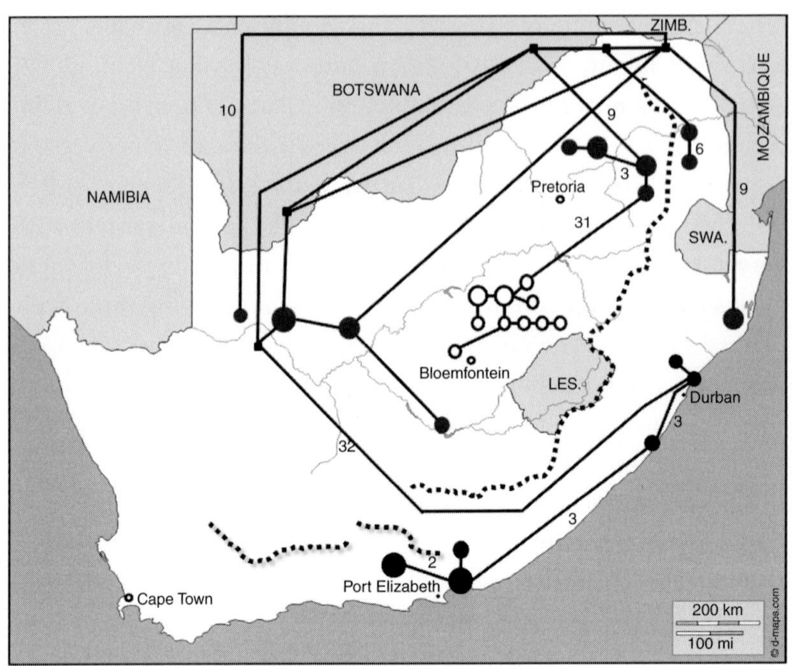

FIGURE 8.1 South African trapping sites. Map created from
http://d-maps.com/carte.php?num_car=11688&lang=en.

Although our results show distinct patterns of genetic structure across the South African landscape, levels of genetic diversity within individual populations were in some cases very low or
absent. In fact, 10 out of the 15 regional localities studied showed
no variation. This fixation for single haplotypes is probably a reflection of the sex-linked dispersal in vervet monkeys. Since dispersal in
cercopithecine primates is largely dependent on male movements,
with female philopatry (Melnick & Pearl, 1987; Pusey & Packer,
1987), it is expected that mtDNA diversity will suggest strong homogeneity within troops. Nevertheless, historical events involving the
expansion and contraction of habitat associated with glacial maxima
could have resulted in confinement of populations to refugia, followed
by genetic drift. Divergence induced by these phenomena would be

amplified by the small effective population size of mtDNA markers compared to nuclear markers (Birky et al., 1983).

We recommend that future work in this field should include a larger number of genetic markers, including nuclear markers (to gauge the male contribution to current patterns of diversity) and markers with adaptive significance in order to determine the selective and evolutionary implications of the genetic differences observed (Turner et al., 2017). Immune-linked (e.g. MHC or TLR gene families) and stress response-linked (e.g., HSP70 gene family) genes could be suitable candidate genes for future analyses of adaptive variation among regionally dispersed populations. Future field sampling should also include more geographic coverage in order to identify specific regions of divergence giving rise to the overall genetic structure observed.

Part IV **Behavioral Ecology**

9 Behavioral Ecology of Savanna Monkeys

Trudy R. Turner, Christopher A. Schmitt, and Jennifer Danzy Cramer

The guenon tribe, Cercopithecini (Butynksi et al., 2013), consists of a large group of closely related but diverse species of long-tailed, medium-sized monkeys found throughout the African tropics. They are currently divided into six genera. Guenons are primarily frugivorous omnivores (Butynski, 1988). Fruits are the most preferred and contested food items, but insects and flowers also are important parts of the diet. Most species are forest-living, inhabiting primary or secondary rainforests in equatorial latitudes of West and Central Africa. Savanna monkeys have the broadest distribution of guenons and, indeed, of all African monkeys (Struhsaker, 1967; Fedigan & Fedigan, 1988; Butynski et al., 2013). They occupy seasonally dry, forest-fringe habitats throughout sub-Saharan Africa and are abundant throughout savannas north and south of the equatorial forests from Senegal to Somalia, from the Nile River in northern Sudan to the tip of South Africa (Lernould, 1988). They may be the world's commonest species of monkeys (Fedigan & Fedigan, 1988). There also are transplanted populations on the islands of St. Kitts, Nevis, and Barbados in the Caribbean (McGuire, 1974; Chapman & Fedigan, 1984).

Due to their broad distribution, savanna monkeys are found in a wide variety of environmental circumstances from semidesert to lowland swamp and montane forest (Butynski, 1988; Rowe, 1996; Butynski et al., 2013; Mittermeier et al., 2013). Although savanna monkeys are most often found in riverine woodlands, they exploit a wide range of habitats, including rainforest fringe, cloud forest, mangrove swamp, and human farms and tourist parks. Savanna monkeys appear to thrive under the energy-poor conditions of marginal and declining habitats as well as the energy-rich conditions provided by human activities, a versatility that reflects their ability to rapidly adapt diet, ranging patterns, and reproduction to changing conditions (Fedigan & Fedigan, 1988). They are flexible and responsive to dynamic environmental conditions such as fire (Enstam, 2007). The limiting factors for suitable habitats are preferable sleeping sites in large trees (Isbell & Young, 1993) and access to water (Hill, 1966; Cheney et al., 1988).

ORIGIN OF SAVANNA MONKEYS

Cercopithecoids first appear in the fossil record 19 million years ago, during the early Miocene predominance of ape species, and achieve prominence during the late Miocene (Leakey, 1988). These early monkeys were small, semiterrestrial frugivores living in semiarid woodland and gallery forest along seasonally flooded streams (Leakey, 1988; Benefit, 1999). Similar habitats are found today in eastern and northern parts of lowland Kenya, desolate landscapes of bare earth dotted by scrubby *Acacia* and *Commiphora* bushes. Narrow strips of gallery forest provide support to troops of baboons and savanna monkeys, following lifeways likely similar to those of ancient monkeys. Early Miocene fossil assemblages and canine sexual dimorphism suggest a social system comprised of multiple males and females where maturing males migrated from their natal group, similar to those of modern-day baboons and savanna monkeys (Benefit, 1999). Monkeys with the distinctive bilopodont dentition of modern-day cercopithecines and colobines first appear in the upper Miocene 5–10 million years ago and undergo adaptive radiations in the Plio-Pleistocene 1–5 million years ago (Pickford, 1988). The Papionini (baboons and macaques) diverged around 7 million years ago when the Sahara separated the ancestral papionins into baboons and mangabeys south of the Sahara and macaques in Africa and Eurasia (Delson, 1992). Guenons continue the trend toward reduction of the muzzle and reduced sexual dimorphism seen in the mangabeys (Kingdon, 1971).

The earliest evidence of guenons appears in the fossil record at about 2.9 million years ago (Leakey, 1988). While fossil information has been found for *Parapapio*, *Theropithecus*, and *Cercopithecoides*, there are no early fossil *Cercopithecus* or *Chlorocebus* finds (see Chapter 3, this volume). The forest-living guenons are thought to have originated in the coastal forests of West Africa and followed the gallery forests eastward in waves of expansion during successive wet and dry periods. During the last 2.3 million years, repeated oscillations between glacials and interglacials have generated as many as 20 cycles of advance and retreat of tropical African forests (Hamilton, 1988). Resulting forest discontinuities are likely to have contributed to the genetic divergence of guenon species dependent on lowland rainforest, but would have limited impact upon species adapted to gallery or savanna woodland (Oates, 1988).

UNDERSTANDING SAVANNA MONKEY BEHAVIOR: SOCIOECOLOGICAL MODELS

The relation of ecology to the evolution of primate social organization has been a long-term focus of primatology (see Janson, 2000). The earliest approaches mapped simple typologies of social systems onto crude habitat categories (Crook, 1966). The goal was to identify the social system best adapted to each habitat, constructing a quasi-evolutionary sequence for the evolution of the human social system. These typologies rapidly become outmoded as burgeoning field studies documented a degree of social diversity not readily subsumed under simple typologies. Subsequent studies dissected social structure into separate components (group size, number of males, etc.) that were tested for correlations with parameters drawn from ecology (resource size and spacing) and morphology (body size, sexual dimorphism) (Milton & May, 1976; Clutton-Brock & Harvey, 1977). Neither approach provided much explanation for why these associations might occur, however, and those that were offered were either blatantly or tacitly group selectionist (Denham, 1971), treating the group as the adaptive unit. Structural features of the environment like food patch size were related to feeding competition and the economy of travel, but these effects were generally expressed in terms of benefits and costs to the group. Sociobiology and behavioral ecology introduced a more strategic perspective that emphasized individual decision-making among alternative actions with varying costs and benefits. Using comparative data to test specific hypotheses, this research explored the role of resource defense (Wrangham, 1980), predator protection (van Schaik, 1983; Terborgh & Janson, 1986), and sexual conflict (Smuts & Smuts, 1993; van Schaik & Kappeler, 1993) in the evolution of primate sociality.

The resulting socioecological model begins with the assumption that females in most species are limited by their access to resources whereas males are limited by their access to females (Trivers, 1972). Therefore, females would be expected to distribute themselves according to the availability of resources and environmental risks whereas males should respond with strategies designed to optimize their reproductive access to females (Emlen & Oring, 1977; Wrangham, 1980). Female sociality is constrained by feeding competition (Wrangham, 1980; van Schaik, 1983; Sterck et al., 1997) and by male attempts to sequester females or harm the offspring of other males (van Schaik, 1996). On this much there is agreement. The debate comes over the relative importance of predation risk (van Schaik, 1983, 1989; Nunn, 1999) and between-group feeding competition (Wrangham, 1980) in determining female sociality.

BEHAVIOR

Group Composition

Guenon groups generally are composed of 10–40 animals (Cords, 1987). Savanna monkeys live in multi-male/multi-female groups of about 5–76 (Kingdon & Largen, 1997). Like most other cercopithecines, females are philopatric, remaining in their natal groups, while males transfer at maturity. Groups are composed of a core of related females and their offspring. Compared with other cercopithecine species, savanna monkey females are less spatially cohesive than other cercopithecine females (Andelman, 1985).

The benefits of larger group size have been demonstrated in vervet monkeys. Among Amboseli vervets, large groups had better-quality and larger home ranges than small groups (Cheney, 1987). Over a ten-year period, a larger group expanded its territory into preferred areas while smaller groups were pushed into more marginal habitats. Home range size was larger in groups with more females relative to the number in other groups. Small groups suffered more incursions by neighboring groups and females in those groups were more aggressive during intergroup encounters. Larger groups experienced somewhat greater female and infant survival than small groups. During a period of declining habitat in Amboseli, larger groups of vervet monkeys were able to take over the home ranges of other groups. When small groups were displaced to areas with fewer trees, they suffered higher mortality (Isbell, 1990). Fusions of vervet monkey groups occurred after they lost their penultimate adult (Isbell, 1991).

Multi-Male Groups and Male Dispersal

In most of the guenons, only one male is found per group, and resident males are generally intolerant of other males. In contrast, savanna monkey, talapoin, and swamp monkey groups are generally multi-male. Among savanna monkeys, some uni-male groups have been reported (Whitten & Turner, 2004), and single peripheral males between groups or attempting to join groups have also been reported (Andelman, 1985).

The factors favoring inter-male tolerance in savanna monkeys have been a focus of attention because they are a multi-male species closely related to the predominantly uni-male clade of guenons. Since both uni-male and multi-male species have multiple males during the breeding season, what is unknown is what allows males in multi-male species to remain throughout the year. Baldellou and Henzi (1992) suggested males might stay in multi-male groups for predator protection, resource or offspring defense, or future mating opportunities. Analyzing data from several vervet monkey populations, Cords (2000) showed that the number of adult males per group was positively correlated with the number of adult females per group. In a separate study, Whitten and Turner (2004) also found that the number of resident males was closely linked with the number of adult females. Isbell and colleagues (2002) did not find evidence that the multi-male organization of vervets could be due to phylogenetic constraints, males forming coalitions to defend females, or predation risk differences between the sexes. Isbell et al. (2002) suggested that males have limited options for dispersal due to the nature of some vervet habitats (i.e., narrow riparian corridors in otherwise inhospitable environments), and due to these limited dispersal options, multiple males are tolerated within groups.

Males disperse from the natal group around the time of sexual maturity, and usually join adjacent groups (Henzi & Lucas, 1980; Cheney & Seyfarth, 1983; Isbell et al., 2002; Isbell & Jaffe, 2013). Of particular note, Cheney and Seyfarth (1983) found that males may disperse with siblings and join groups where others from their group had previously migrated. Male size may be a factor in this pattern – sexual dimorphism in savanna monkeys is not as extreme as in other terrestrial cercopithecines. Having an ally may allow for coalition formation and increased reproductive access.

Males have a dominance hierarchy and rank is earned through competition and considerable aggression with both resident males and females (Cheney, 1983; Raleigh & McGuire, 1989). Following dispersal, sons of both high- and low-ranking mothers are equally likely to achieve dominance status in their new groups (Cheney & Seyfarth, 1988).

Intragroup Interactions

Interactions between females and males are infrequent in guenon groups. Struhsaker (1967) noted that more intragroup agonism was found in smaller groups. Affiliative interactions are primarily between females, with low levels of observed aggression among females (Whitten, 1983). Females are likely to be aggressive toward adult males (Rowell, 1988). Individuals and matrilines form alliances during aggressive intergroup and intragroup interactions (Cheney, 1983). Aggressive interactions can extend beyond two participants to involve kin. When two vervets are in conflict or in reconciliation, they are also in conflict or reconciliation with the other individual's kin (Cheney & Seyfarth, 1989).

Females are known to form alliances against males (Andelman, 1985). Cheney (1983) suggested that kin selection plays a role in female alliances against males of all ranks. Females help maintain order and prevent male aggression toward young (Kingdon, 1997). Females prefer higher-ranking alliance partners to kin or lower-ranking females (Cheney, 1983). Although higher-ranking females are unlikely to reciprocate alliances for lower-ranking females, lower-ranking females may earn more tolerance at feeding sites or receive less aggression after allying with higher-ranking females (Cheney, 1983).

Dominant males differ in behavior, having a higher number of affiliative interactions with females (McGuire & Raleigh, 1985) and higher rates of aggression (Raleigh & McGuire, 1990). Alpha males are more affiliative to young group members than subordinate males, and females are more tolerant of subordinate males who behave affiliatively to their offspring (Hector et al., 1989). Although savanna monkey males do not provide paternal care of offspring, males appear to behave affiliatively toward young monkeys depending on whether the offspring's mother is watching the interaction (Hector et al., 1989).

Intergroup Dynamics

In contrast to the low-key interactions within guenon groups, interactions between savanna monkey groups are marked by chases and physical aggression. Savanna monkeys are territorial with home ranges of 13–178 ha (Kingdon, 1997) depending on the quality of the habitat (Struhsaker, 1967). In South Africa, troops have well-defined ranges with little overlap (DeMoor & Steffens, 1972; Barrett et al., 2010). Females are vigorous participants in boundary disputes, whereas resident males rarely participate in intergroup encounters but are intolerant of other males (Rowell, 1988). Vervets are an exception in that both males and females participate. Vervets are known to move into other vervet territories and supplant resident groups (Isbell et al., 1990).

Communication

The guenons might be said to be the shyer cousins of the papionins, the other subfamily of the cercopithecines. In the papionins, rank relationships are especially well defined and are maintained with an elaborate set of overt facial expressions, gestures, and calls (Rowell, 1988). In contrast, the guenons negotiate their social relationships with a far more subtle set of sly glances and postural shifts, a Noh counterpoint to the stylized Kabuki dramas of the papionins.

Drab gray and green coat colors add to the understated persona of guenons, but their faces and genitalia can be quite extravagant. Bright facial patterns highlight ritualized head movements that accentuate eye avoidance used as a submissive signal (Kingdon, 1992). In the arboreal guenons, these displays have been replaced by brightly colored facial patterns that accentuate ritualized stare-threat and aggression-dampening eye-avoidance signals (Kingdon, 1988). Savanna monkeys have black faces and grayish green pelage. Taxa can be defined by different morphological characteristics, including the presence of a frontal band across the face, the presence and color of facial whiskers, the greenness of the pelage, the presence of a yellow or black tail tip, and the presence of a red patch at the base of the tail (Hill, 1966).

Male Color Communication

A specialized signal repertoire helps to mediate coexistence in savanna monkeys (Henzi, 1985; Cords, 2000). Males use brightly colored genitalia to accentuate behavioral displays (Gerald, 2001; Cramer et al., 2013). Only a few primate genera have blue sexual skin: savanna monkeys, talapoin monkeys, patas monkeys, some guenon species, and mandrills. Hill (1966) described in detail how scrotal skin color differs across geographic populations. In West Africa, *Ch. sabaeus* have white or faint blue scrota. In Central and East Africa, both *Ch. tantalus* and *Ch. aethiops* were described as having sky blue scrota (Hill, 1966). In southern Africa, *Ch. pygerythrus* are described as more vivid, with scrotal color ranging from blue to violet (Hill, 1966). This geographic variation between populations becomes more interesting as later researchers described how within a population scrotal color also varies among individuals (Henzi, 1985; Gerald, 1999).

Recent developments in digital photography have brought new opportunities for researchers studying color. Catching up to the rich body of literature on avian color and the objective field methods used by ornithologists (Montgomerie, 2006), over the last decade primatologists have been using digital photography to objectively measure and study primate color with respect to what we know of visual perception among trichromatic primates (Higham et al., 2010). For blue skin specifically, researchers measure the color (chromatic) and luminance (achromatic) channels (Stevens et al., 2009; Higham et al., 2010). Color refers to the perceived blueness of the color, and luminance refers to the perceived lightness of the color (Cramer et al., 2013).

Until recently, studies of savanna monkey scrotal color were largely anecdotal. Brain (1965) and Gartlan and Brain (1968) noted that, in the wild, scrotal color pales during physiological stress or following stressful social events. In a study examining the genital signals of vervets, Henzi (1985) found pale males are not necessarily more subordinate than darker males. This finding is difficult to interpret because color was not objectively defined and measured as part of the study. In a later study examining seasonal changes to vervet hair and skin, Isbell (1995) found a nonsignificant trend between darker males and higher ranks. This study was the first to begin examining vervet scrota in a measurable way, with Isbell assigning males to one of four color categories. Brain's early suggestion that vervet color changes with aggression or illness persists with anecdotal support. Among captive, wild-born vervets at the Vervet Research Foundation in South Africa, this is also the case (Du Toit, personal communication). Fairbanks (personal communication) suggests that scrotal color pales after wild males are trapped, an effect not observed during a large-scale trapping project involving hundreds of male vervets (Cramer et al., 2013). Cramer et al. (2014) examined the relationship between a hormonal measure of stress, fecal cortisol, to examine the potential link between stress and color. Males with lower fecal cortisol concentrations were bluer and lighter in color (Cramer et al., 2014).

A key first step in understanding the range of variation in savanna monkey scrotal color is to measure it across widespread, wild populations. Examining scrotal color across Caribbean and African populations, Cramer and colleagues (2013) used digital photography and color analysis software to analyze color data from 348 males. Fitting with the phenotypic variation described in Hill's sub-specific diagnostic features (1966), Cramer et al. (2013) found that scrotal color significantly varied between the two most distant taxa, *Ch. sabaeus* and *Ch. pygerythrus*. *Ch. pygerythrus* are significantly bluer than *Ch. sabaeus*. Cramer and colleagues (2013) also found significant variation within each taxa. Similarly, Caribbean *Ch. sabaeus* adults were significantly lighter than West African *Ch. sabaeus* adults. These site-specific color patterns suggest the potential for environmental factors, genetic relatedness, or some other cause in determining scrotal color.

Color also significantly varied across age categories (Cramer et al., 2013). In the more vivid taxa, *Ch. pygerythrus*, males became bluer with age. In the lighter, sometimes achromatic taxa, *Ch. sabaeus*, males become less blue with age. For both taxa, infant and juvenile males had dull blue scrota, diverging as subadults and adults when *Ch. pygerythrus* become vivid and *Ch. sabaeus* become light or achromatic. In these two taxa, color changes across the life span suggested color may function as an age-related signal. Among gelada baboons, chest patch redness is an age-related signal of sexual maturation (Bergman & Beehner, 2008). Blue scrotal color could function similarly in savanna monkeys. The blue sexual skin of savanna monkeys results from the nanostructure of dermal collagen (Prum & Torres, 2004), and the mechanism by which color change happens across a life span is unclear.

Color is an important component in savanna monkey behavior. Field studies of *Ch. pygerythrus* suggest that scrotal color plays a role in intrasexual communication (Struhsaker, 1967; Durham, 1969; Henzi, 1985). Adult males engage in a red, white, and blue display where the contrasting colors of their perineal region are prominently presented to conspecifics (Struhsaker, 1967). These highly contrasting colors of the perineal region are an important part of the display (Struhsaker, 1967; Durham, 1969). This behavior is particularly common to aggressive interactions during the mating season (Struhsaker, 1967). Of the five primate taxa with blue sexual skin, only savanna monkey males are described as using this color in communication during the red, white, and blue display (Dixson, 1998). Body size and canine size are known to be important in male–male competition among some primate species (Dixson, 1998). Examining blue color as a potentially sexually selected trait, Cramer and colleagues (2013) identified relationships between male color and morphological traits. Among *Ch. pygerythrus*, bluer adult males had lower body weights. Adult *Ch. pygerythrus* with longer canines were lighter. Contrasting with Gerald's (1999) study of captive *Ch. sabaeus* that found that "dark" phenotype males were significantly taller and had larger absolute testes size, Cramer and colleagues did not find a relationship between color and these morphological variables. Instead, Cramer et al. (2013) found that among *Ch. sabaeus*, lighter adult males had longer body lengths. Additional study is needed to identify the mechanisms underlying the relationship between color and morphology and how it differs between captive and wild populations.

Though the study of the communicative role of primate coloration has been increasingly popular in recent decades (see Semple & Higham, 2013), very few studies have examined scrotal coloration in captive and wild savanna monkey populations. In a series of studies, Gerald (1999, 2001) and colleagues (2010) used captive populations from Caribbean *Ch. sabaeus* stock to experimentally alter the color of males and study the behavioral outcomes. These studies suggested that scrotal color played an important role in mediating conflict and that altered males received more aggression than unaltered males (Gerald, 2001; Gerald et al., 2010). The results of these studies are difficult to interpret because these captive *Ch. sabaeus* are mostly achromatic, with only some males having the light blue or blue coloration described by Hill (1966) and others as diagnostic to African savanna monkeys. Experimentally altering males beyond their natural condition may have a significant impact on the behavior of animals (Hill, 2006).

In a behavioral study of ten males in two troops, Cramer (2013) and colleagues (2014) found no significant link between male color and male rank. This finding supports Henzi's (1985) suggestion that color may not reliably predict rank. Based on Gerald's (2001) finding that color predicted aggressive outcomes between captive *Ch. sabaeus* males, Cramer and colleagues measured rates of aggressive behavior among wild *Ch. pygerythrus* males. Among the wild *Ch. pygerythrus* in this study, no relationship between aggressive behavior and color was found. These opposite findings point to the need for more behavioral research in both taxa to identify how captive or wild status might have an impact on the predicted relationship between color and aggressive behavior, or why there might be taxonomic differences in this relationship. The results of these experimental studies on scrotal color likely have limited application in understanding scrotal color variation among wild African populations of *Ch. pygerythrus* as well as *Ch. sabaeus*.

Reproduction

Observation of savanna monkey mating behavior is rare (Struhsaker, 1967). Females lack ovulatory cues during the mating season and typically begin mating several months before cycling (Andelman, 1985). Unlike many cercopithecines, savanna monkey females do not display sexual swellings or sexual skin color changes to communicate changes in reproductive status. Males initiate most copulations (Andelman, 1987). Struhsaker's early study at Amboseli (1967) noted female solicitation behaviors, but Andelman's later work at Amboseli (1985) did not report female solicitation behaviors. After copulation, females sometimes leave the male entirely (Andelman, 1985) rather than have long-term relationships or consortships as seen in some other cercopithecine species (Dixson, 1998). Older, dominant males had the highest frequency of copulations with females, although females actively reject most copulation attempts by high-ranking males (Andelman, 1985).

Dominant females may have priority-of-access to preferred males (Rapkin et al., 1995). A field study in Kenya found that females typically begin cycling after the mating season has begun (Andelman, 1985). Andelman's study (1985) suggests that male age and dominance status are the most important factors in female mate choice.

Savanna monkey females continue mating well into the first half of pregnancy (Andelman, 1985), a pattern that is also typical of rhesus macaques (Bielert et al., 1976) and may help females confuse paternity. Andelman (1987) suggested that concealed ovulation in savanna monkeys helps increase male–female bonds, increase paternal care, and decrease male–male competition.

Females have an interbirth interval of one to two years (Lee, 1984; Isbell & Enstam Jaffe, 2013). In a field study of three vervet troops, Turner and colleagues (1987) found that high-ranking females reproduced more frequently, and there was significant variance in the number of live births between observation years. Shorter interbirth intervals and more synchronous breeding may provide vervet males with more mating opportunities than in forest guenons. However, patas monkeys have interbirth intervals and breeding synchrony similar to those of vervets and even larger numbers of females per group, yet do not live in multi-male groups.

Among females, allomothering appears to play an important role in the behavior of young primiparous females and allows mothers to spend more time on other activities (Fairbanks, 1990).

The bond between mothers and daughters seems to be important for reproductive success. Females without a living mother do not reproduce as successfully as females with a mother in the group (Fairbanks & McGuire, 1986). Females with living kin in the troop tend to have higher rates of reproductive success than females without close matrilineal support (Fairbanks & McGuire, 1986).

Rainfall is likely to be a major factor in the timing of guenon reproductive seasons (Butynski, 1988). Most guenon species, including savanna monkeys, conceive during the dry season months (in East Africa) of July–September and give birth during the rainy season months of December–February, the period of greatest food abundance (Butynski, 1988). Vervet monkey infants begin to forage for themselves at around three months of age, a period that also coincides with the period of most abundant food in areas with low rainfall ($\leq$101 cm/year; Butynski, 1988). In areas where greater rainfall provides more sustained periods of food availability, savanna monkey infants are born at the onset or during the dry season, a strategy that may protect infants from diseases associated with wet conditions (Gautier & Gautier-Hion, 1968). However, an alternative hypothesis must be considered: that the proximate cues that time mating and conception are conserved across guenon species and are largely responsible for the timing of births across guenon species. This possibility is suggested by the similarity of birth months across guenons in very different habitats and the conservation of similar gestation length.

Predation

While primate fieldworkers have viewed predation as a powerful selective force on primate behavior, actual evidence of predation is hard to find. Early descriptions of savanna baboons depicted predation as an ever-present threat that maintained sociality and molded the structure, daily progressions, behaviors, and sleeping patterns of the primate group (Washburn & DeVore, 1961). In East Africa, any moment may bring an encounter with a lion, leopard, hyena, cheetah, or pack of wild dogs. A primate quickly becomes attuned to the sounds and silhouettes of the environment and the changes that signal danger: an impala's snort, a monkey's gaze, the rustle of a bush, a sudden stillness. There is a greater sense of security in the presence of the group – no alarm calls means no predator – and the primate observer begins to depend on the primates' invariant ability to spot the predator first. This collective vigilance has been presumed to be a major benefit of social life for primates (Isbell, 1994a) and could be one reason why savanna monkeys live in multi-male/multi-female groups (Isbell et al., 2002). More eyes and ears increase the likelihood of spotting a predator so that group members can devote more attention to foraging, but these benefits may diminish once a minimal size is reached (Janson, 2000). Recently, researchers (Willems & Hill, 2009a, 2009b; Coleman & Hill, 2014) have described ranging behavior and predation risk as part of a "landscape of fear." They suggest that prey animals, such as vervets, modify their behavior in response to variations in predation risks. They were able to quantify this hypothesis by observing alarm calls and animals' responses in home range use to these calls. They suggest that predation risk is a key driver of behavior with direct and indirect costs to fitness. Direct costs are, of course, loss of life. Indirect costs are the behaviors that are engaged in that reduce risk, producing trade-offs between energy intake and predation avoidance.

The major predator of savanna monkeys is probably humans. Animals are hunted and killed because they raid human habitations. Besides humans and dogs, lions, leopards, cheetahs, hyenas, jackals, servals, caracals, eagles, rock pythons, and crocodiles all prey upon vervets. Other species of primates, including chimps and baboons, also prey upon them (Pruetz & Bertolani, 2007; Butynski et al., 2013). Cheney et al. (1988) estimated that predation accounted for 67 percent of all vervet deaths at Amboseli. Both Cheney et al. (1988) and Isbell et al. (2009) indicate that only between 15 and 27 percent of vervets make it to adulthood. The amount of time vervets spend scanning the environment could be related to this threat of predation.

Vocal Communication

Struhsaker's classic monograph (1967) details the complex vocal communication system of vervets. Vervets have a wide vocal repertoire and give specific responses to their aerial and ground predators (Cheney et al., 1988; Isbell, 1990; Isbell & Enstam, 2002). Vocalizations are specific to ambush and ground predators and alert group members to take appropriate action by mobbing the predator or seeking shelter (Cheney & Seyfarth, 1998). In addition to their own vocal responses to predators, vervets react to the predator calls of other species (Cheney & Seyfarth, 1998).

In a series of studies, Cheney and Seyfarth examined vervet communication in considerable detail, highlighting the animal's cognitive abilities. Their work identified how predator-specific calls were learned, not imitated, and how variations made by just-learning juveniles are understood by adult group members (Cheney & Seyfarth, 1980). Designing playback experiments to help determine what vervets know and assume based on vocalizations, Cheney and Seyfarth (1980, 1982) found that female vervets know the screams of individual juveniles and associate those screams with the juvenile's mother, presenting an opportunity for the female to consider different behavioral responses and the benefits and consequences. The cognitive ability to use past experiences to predict future interactions is important for balancing cooperation and competition (Cheney & Seyfarth, 1986). Among vervets, Cheney and Seyfarth found that redirected aggression and reconciliation behaviors are never random, but instead stem from previous interactions (Cheney & Seyfarth, 1990b). Predicting risk in interactions may be particularly important for mothers and infants, and experimental work by Fairbanks and McGuire (1993) found this to be the case when unknown males were introduced to mother–infant pairs. Similarly, Hector and colleagues (1989) found that males behaved differently with juveniles and infants if they knew that their mother could see the interaction. This suggests that males are attuned to the consequences of their actions and behave differently depending on the consequence risk.

Cognition

Among primates, sociality is an important correlate of cognition (Essock-Vitale & Seyfarth, 1987). Primates must learn and store information not only applicable to current circumstances, but to broader, future circumstances (Cheney & Seyfarth, 1990b). Primates have to keep track of complex social relationships with kin and non-kin and internalize and make judgments about others (Cheney & Seyfarth, 1990a). Studies of cognition in wild savanna monkeys really stem from Cheney and Seyfarth's landmark studies on vocalization and communication. Their fieldwork also highlighted kin recognition among vervets and the complex decision-making and trade-offs in vervet social relationships.

Vervets use mental mapping to keep track of conspecifics (Noe & Laporte, 2014) as well as food sources (Cramer & Gallistel, 1997). Tracking family members is important (Seyfarth & Cheney, 1984), as is keeping track of the group as a whole. Noe and Laporte (2014) found that vervets can track group members in space, silently or with vocalizations. Keeping track of group members, with or without vocalizations, helps with group cohesion, providing protection against predators and other groups. Vocalizations also contribute to mental mapping of other troops (Cheney & Seyfarth, 1982).

Though savanna monkeys do not typically use tools, there is also some research suggesting that they have the cognitive abilities to use them in limited ways (Santos et al., 2006), especially in human environments (van de Waal & Bshary, 2010). In the wild, cognitive studies of vervet monkeys continue, with researchers studying social learning (van de Waal, this volume) as well as novelty-seeking (Blaszczyk, this volume).

Savanna monkeys are an incredibly plastic primate species adapted to many different habitats throughout sub-Saharan Africa and the Caribbean. They demonstrate flexibility when faced with severe changes to the environment such as drought or fire (Isbell, 1995; Jaffe & Isbell, 2009). They also demonstrate a high degree of flexibility in highly fragmented and disturbed environments, sometimes opportunistically coexisting with humans. Their environmental flexibility as well as their complex cognitive abilities and communication can give us important insights into early common ancestors and human evolution.

10 Socioecology of Vervet Monkeys

Patricia Whitten

A soft "waaa" wafted across the shaded recesses of the acacia grove. A female shifted, roused from her nap, and looked around. LF's infant lost-called again, and a juvenile sidled up to the infant. She began to groom, running her fingers studiously through the infant's baby-fine coat. The infant sat rigidly, staring out beyond the grove. The craggy outline of Lowa Mara shimmered over the spare yellow grass of the dusty plain. LF was nowhere to be seen. She was in fact more than a kilometer upstream, in another grove of acacia trees. LF, a middle-ranking female in LM troop, had stayed behind to feed with another female as the rest of the troop moved on. Following juvenile playmates, the infant had gotten separated from its mother. The afternoon shadows grew longer, and still LF did not appear. Now a lion rested near her feeding spot, an effective deterrent to further travel. The infant spent the night without its mother, finding what comfort it could huddled with a juvenile. The next morning, LF and her companions traveled rapidly west, pausing only briefly to survey open spots and then dashing quickly across them. As LF entered the grove, her infant ran toward her. The two embraced, and the infant began to suckle.

This episode was only one of many confusing events resulting from group fissions during Whitten's fieldwork in Samburu. Beginning in August of the long dry season, the two main study groups, RR and LM, divided into subgroups for the next three months. Surveys in other parts of the Reserve suggested that fissions were widespread. In each of the study groups, a group of low-ranking females and their dependent offspring separated from the parent group, remaining in a small portion of the home range while the rest of the group continued to make their usual circuit through the home range. The first splinter group formed when LN, the eighth-ranking of nine LM females, and her two offspring took up residence at the eastern end of the home range. Shortly thereafter, LF, PN, and DN, the sixth-, seventh-, and ninth-ranking females, began to stay at the center of the home range. NT and RE, the seventh- and eighth-ranking of eight RR females, remained at the western end of the home range, joined later by the sixth-ranking female, BT, and briefly by the fifth, CT. Initially, relations between splinter groups and parent groups were congenial. Members of the parent group groomed and fed with splinter group females when they happened upon them, and occasionally an adult or juvenile from the parent group joined the splinter group for a few hours or days. Gradually, however, relations between the subgroups became more tense. About a month after the first separation, LM members began to chase and harass splinter group females encountered while traveling through the home range, much as they would members of an alien group. Similar hostilities occurred in RR troop two months later. The attacks were generally carried out by the dominant male in the group, who generally preceded the group into feeding areas, searching for intruders, but females took part when splinter group females were found feeding in prized *Acacia tortilis* trees. Female and sometimes male members of the splinter group rushed to defend females who were attacked, but generally ended up withdrawing. After a while, splinter groups avoided all contact with the parent group, fading into the underbrush whenever they approached. The splinter groups were effectively separate groups. Females who were once compatriots were now adversaries.

ECOLOGICAL BASIS OF FISSIONS

What caused these striking changes in group composition and allegiance? Ecological changes appeared to be the precipitating factors. The events occurred during a disruption in seasonal rainfall patterns. Like most of East Africa, Samburu is characterized by four distinct seasons. The annual cycle begins in October when rains brought by the Indian monsoons awaken the landscape from the long drought of the preceding dry season. These are the short rains, brief showers that darken the afternoon skies and produce a carpet of fresh grass across the bare plains. New vervet mothers busy themselves with harvesting this crop, plucking handful after handful of the freshest green blades. By December, young forbs have sprouted beneath the acacias, and the vervets begin to harvest their tiny flowers. As the rains come to an end in January, the *Acacia tortilis* trees turn snow white with a profusion of blossoms, which the vervets devour from dawn to dusk through February. Beginning in March, the drenching rains of the long wet season turn roads to rivers and fill long-dry stream beds. Now the *Acacia elatior* trees of the riverine forests begin to bloom in synchrony, and the vervets disappear for hours foraging within their great crowns. *Acacia tortilis* pods appear as the rainy season draws to a close, and the ripe seeds become a favored food in the ensuing long dry season, along with ripe fruit from several species. Late in the dry season, *Acacia tortilis* flowers again, although less profusely, and it again becomes an important food.

Changes in rainfall altered this phenological cycle and intensified female feeding competition. In the year of the fissions, the long rains began and ended earlier than in previous years, and the short rains were late, producing an extended dry spell lasting from February until November. These differences in rainfall patterns delayed the flowering and fruiting of the major dry season resources. In RR, females were forced to eat the fruit of one favored plant, *Aphania senegalensis*, before they were ripe, normally a strategy employed only by low-ranking females. Food and energy intake were substantially less in the two months preceding the fissions than in the same months in the previous year. The deficits were primarily due to reductions in the intake of *Acacia tortilis* flowers and seeds. These deficits affected high-ranking as well as low-ranking females. Rank was not correlated with food intake in either group in June–July leading up to the fissions, even though RR food intake increased with higher rank in the year before the fissions. However, the females who first initiated fissions – bottom-ranking NT and RE in RR and LN in LM – had very low energy intake in the months before the fissions.

CONSEQUENCES OF FISSIONS

The effects of the fissions on energy intake varied depending on the location of the splinter groups. By staying in one spot, the splinter groups had more opportunity to harvest the local resources and avoided competition from higher-ranking females. Supplant rates declined during the fission period, particularly for low-ranking females. In LM, the LF splinter group spent its time in one of the major groves of *Acacia tortilis* in LM's home range. LN isolated herself and her offspring in an area with several *Acacia tortilis* trees. During the fission period, these females not only gained more energy from *Acacia tortilis* than in the months just prior to the fissions, but also gained significantly more energy than the high-ranking females who remained in the parent group. In RR, on the other hand, the NT subgroup stayed near the edge of the home range in an area where there were only two *Acacia tortilis* trees. During the fission period, these splinter group females gained significantly less energy from *Acacia tortilis* flowers than females in the parent group. In fact, at some points, they appeared to subsist mainly on the long green thorns of *Acacia elatior*, and NT could be seen each morning regurgitating and re-ingesting their viscous green residue.

There were other costs as well. Their smaller size may have forced both parent and splinter groups to be more vigilant toward predators. Alarm call rates can be used as an index of perceived predator risk (Altmann & Altmann, 1970; Cowlishaw, 1997). In both troops, alarm calls were more frequent during the months of the fission, which might indicate that females became more vigilant when party size was reduced by the fissions. However, alarm calls were also more frequent in the pre- and post-fission months than in the previous year, suggesting that the higher frequencies may reflect other factors. Cowlishaw found that female alarm call rate declined with increasing group size in Namibian baboons (Cowlishaw, 1997), but Isbell (1991) and Stanford (1995) did not find a relation of alarm call rate to group size.

The small splinter groups also were more vulnerable to neighboring groups, which almost invariably displaced them. On one occasion, however, the three females and yearling of the RR splinter group managed to repeatedly chase off the single male, OE, from a neighboring group. At midday, the parent group, ranging parallel to the splinter group on the other side of the river, watched the splinter group's interactions with OE and began threatening and giving intergroup contact calls. GP, a male who had previously associated closely with the splinter group females, crossed the river and joined them in repulsing OE. Later that afternoon, GP chased OE into the center of the parent group, where he was threatened and chased.

The latter observation suggests that, in spite of aggression between subgroups, some recognition of original group membership was maintained. In fact, with the onset of the short rains in mid-November, all splinter groups began to rejoin their parent groups. The rains carpeted the ground with fresh grass, a resource that tended to minimize feeding differences among females. Splinter group females began to feed and sleep with the parent group. At first, splinter group members were attacked by the dominant male, but soon they were traveling peaceably with the parent group. Some parent group females initiated grooming bouts with splinter group females, although they still remained rather peripheral. However, by early January, the subgroups formed again.

EXPLAINING SOCIAL SYSTEMS

These events not only show that ecology can profoundly affect group composition, but also raise some interesting questions about the determinants of primate sociality. How did the Samburu vervets slip so easily between confrontation and fidelity? How do ecology and social interaction affect sociality? What is the basis of group allegiance? What maintains group cohesion and composition?

Sociality is a fundamental part of existence for most primates. Even the "solitary" nocturnal prosimians sleep with conspecifics and coordinate foraging pathways with neighbors (Bearder, 1987). Among mammals, primates are notable for the complexity of their social systems (Wilson, 1975; Eisenberg, 1981). This complex social environment has become a key ingredient in the normal milieu for primate development. The social environment has been an important selective factor in primate evolution. Group-living primates must process a complex and rapidly changing array of visual, auditory, and olfactory information about identity, social attributes, intentions, behaviors, and past actions of multiple individuals. These demands constituted a crucial selective force in the expansion of the primate neocortex (Barton & Harvey, 2000) and define the order Primates.

11 Biological Complexity in Primate Sociality and Health

Brandi T. Wren

It is well known that primates are particularly social animals (Smuts et al., 1987). By pairing our capacities for extreme sociality and a highly developed intelligence, humans have created technologies and communication systems that allow individuals around the globe to connect within seconds and travel with ease. These adaptations facilitate not only the sharing of ideas around the world, but also the transmission of disease. The extreme example of HIV illustrates this: an infectious disease with roots in nonhuman primates that has infected humans around the world. The act of being social, however, can affect an individual in myriad ways, from enhancing health to degrading it. There is a robust literature that confirms that being highly social is beneficial to one's health. In humans, having more and frequent social contacts is associated with living a longer, healthier life (Berkman, 1984) and giving birth to healthier infants (Collins et al., 1993). For baboons, being more social also means giving birth to more infants (Silk et al., 2003; Silk, 2007a, 2007b).

It is relatively easy for us to imagine the degree of sociality that some modern human innovations required, as well as how they might influence disease ecology not just within a primate species, but also among them. What about primates foraging in the wild? Are the costs and benefits of being social the same for them? And what does this mean for the evolution of primate social behavior? Behavioral ecology theory predicts that an adaptation is selected for when the costs are exceeded by the benefits. Like many adaptations, sociality involves trade-offs. I argue that the benefits of sociality for primates – including the vervet monkeys (*Chlorocebus pygerythrus*) I study – outweigh the costs, and that this is especially true when we consider the costs and benefits of parasite transmission, stress, and physical touch.

Social grooming is a major component of nonhuman primate sociality. Recognizing this, I focused on social grooming as a mechanism for studying the evolution of primate sociality as well as the influence of social contact on health. I wanted to know how grooming might influence health via transmission of microbes and to explore what this meant for the evolution of primate social behavior. While social grooming can be beneficial due to the removal of ectoparasites (Akinyi et al., 2013), it can also be detrimental because of the transmission of other types of parasites and pathogens (Caillaud et al., 2006). I looked at infections with gastrointestinal parasites because they are common, widespread, and had been overlooked in previous studies of social grooming. I hypothesized that social grooming might transmit a multitude of endoparasites, despite providing protection against ectoparasites.

In 2009, supported by research grants from the Wenner-Gren and Leakey Foundations, I headed to South Africa to study the behavioral ecology of primate–parasite interactions in wild vervet monkeys (*Chlorocebus pygerythrus*) (Wren, 2013). Broadly, I wanted to know whether being more social was associated with greater measures of infection with certain gastrointestinal parasites. Specifically, I wanted to know whether having more grooming partners was associated with infection with a greater number of parasite species (a measure termed "parasite richness" by parasitologists). I also planned to examine whether the amount of time that individuals spend grooming with others was associated with parasite richness. I expected that being more social – either through grooming for longer times or with more individuals – would lead to more infections with certain types of parasites – specifically, those that are directly infective from host to host.

I conducted fieldwork at Loskop Dam Nature Reserve from 2009 to 2010, following three groups of vervets that were already habituated (Barrett et al., 2010; van de Waal et al., 2010) – the Bay,

Donga, and Blesbok groups. Vervet group sizes are fairly typical at Loskop compared to the surrounding area, which vary from roughly 10 to 25 individuals (personal observations). Sizes of the study groups at Loskop ranged from 13 to 20 individuals, fluctuating due to individual migrations, deaths, and births. Home ranges for the study groups converged such that the edges of the Blesbok and Donga home ranges overlapped, as well as the edges of the Donga and Bay groups. The Blesbok group was dominant to the Donga group during this study, displacing them in 100 percent of encounters. During the following field season (2011), I observed the migration of two male subadults from the Blesbok group to the Donga group. Because both of the subadult males were offspring of the dominant female in the Blesbok group, this drastically altered group dynamics. Following these transfers, the Blesbok group always tolerated the Donga group when encountering them at the edge of their home range (in preparation).

All three study groups' ranges were within the boundaries of Loskop Dam Nature Reserve in Mpumalanga Province. The reserve contains approximately 23,000 ha of land as well as the 1847-ha reservoir created by the Loskop Dam itself. Loskop consists of mountainous bushveld surrounding the Olifants River valley, and has a highly seasonal climate of rainy, hot summers and dry, cool winters. Barrett and colleagues noted that the Donga's home range is broadly characterized as *Acacia caffra–Rhus leptodictya* woodland, and further described it as containing *Lippia javanica–Loudetia simplex* shrubland, *Acacia nilotica–Acacia caffra* woodland, and *Olea europaea africana–Rhus leptodictya* woodland (Barrett et al., 2010). The Blesbok group ranges just west of the Donga group, and the Bay range is located to the east.

Almost half (42 percent) of my study groups' diets consisted of fruit. The most commonly ingested fruit species were thorny elm (*Chaetachme aristata*), Transvaal red milkwood (*Mimusops zeyheri*), wild olive (*Olea europaea*), common wild fig and/or broomcluster fig (*Ficus* spp.), white stinkwood (*Celtis africana*), red ivorywood

(*Berchemia zeyheri*), blue guarri (*Euclea crispa*), and mountain karee and/or common wild currant (*Rhus* spp.). This does not include fallen fruit that was taken from the ground while foraging terrestrially. Seeds made up just over 10 percent of the diet of Loskop vervets. Monkeys commonly ingested seeds from the large-fruited bushwillow (*Combretum zeyheri*) and a variety of acacia species (*Acacia* spp.). These seeds were usually masticated to the point where germination following defecation would be impossible, although the smaller seeds inside fruits that were ingested were often seen whole in the feces, with germination still possible. Monkeys were also observed eating mushrooms, assorted vertebrates and invertebrates, soil, a variety of terrestrial herbaceous vegetation, flowers, shoots, lichen, and gum.

Vervet monkeys provide an ideal model for studies on social grooming, particularly when looking at individual variation. Vervets are known to exhibit individual variation in grooming behaviors and to adjust their social grooming behaviors according to the situation (Seyfarth, 1980; Isbell & Young, 1993). As I will discuss below, Loskop vervets appear to be similar to other populations of vervets, as well as primates in general, when it comes to how they choose to spend their time socializing with others. Thus, vervets could serve as a useful model for understanding the evolution of sociality in primates overall (Seyfarth, 1977).

Overall, Loskop vervets spent an average of 28 percent of their non-sleeping day in physical contact with other monkeys, with 9 percent of that time spent grooming with conspecifics. Individuals ranged from 0 to 100 percent of their time in non-grooming direct contact with a conspecific and from 0 to 39 percent of their time in grooming with a conspecific. The individuals spending unusually large amounts of time in contact with others were most often infants who were in contact with their mothers constantly throughout the day or the mothers carrying those infants.

As expected, female vervets at Loskop were more social than males. This is a common trend among both humans (Dunbar & Spoors, 1995) and nonhuman primates (Mitchell & Tokunaga, 1976).

In general, female vervets at Loskop spent roughly 10 percent of their time grooming others, while males spent less than 2 percent of their time grooming others. Interestingly, male and female vervets both spent about the same amount of time each day being groomed (3.2 and 4.6 percent, respectively).

Loskop vervets averaged 4.2 grooming partners per individual for the whole study sample, with a range of 0 to 13. After excluding infants from the analyses, females had more grooming partners than males, averaging 6.9 partners to males' 3.2. These results are similar to those reported for human social networks. Studies suggest that most humans have anywhere from two to five close friends (Dunbar & Spoors, 1995; McPherson et al., 2006).

Vervets at Loskop were infected with a variety of gastrointestinal parasites, most of which were helminths (Wren et al., 2015). Helminths included whipworm (*Trichuris trichiura*), hookworm, stomach worm (order Spirurida), nodular worms (*Oesophagostomum* spp.), and threadworm (*Strongyloides* spp.). Aside from stomach worm, the other helminths are most commonly contracted via skin penetration by infective larvae in contaminated soil. Because these parasites are not infective upon release from the body of a host, we did not expect that physical contact with others would be associated with their transmission. Stomach worm is typically contracted from ingesting intermediate hosts like crickets or other arthropods, so also could provide only limited insight into links between social behavior and infection.

Loskop vervets were only infected with one protozoa, *Entamoeba coli*, but almost the entire population was infected (92 percent). This made it impossible to tease out whether variation in individual behavior was associated with infection status for protozoan species. This was an obvious blow to the project, because it meant I could not test the predictions I had hoped to test. I did stumble across some surprising results, though.

The vervets at Loskop who were infected with hookworm also had significantly more grooming partners than those who were not infected (Wren et al., 2016). This was unexpected because hookworm is not typically considered to be infective between hosts. Hookworms, like many helminth parasites, are released from the host's body in a larval stage that is not capable of infecting another host. Because of this, it was not expected that grooming would increase exposure to hookworm.

The problem with this perspective is that it does not account for the infective stages of larvae that monkeys likely pick up from the soil while on the ground. Vervets at Loskop spend a great deal of time on the ground and could easily come into contact with infective stages of larvae in the soil. It could be possible that infective stages of larvae are left on the hands of the groomer, and then are ingested later while feeding. Infective stages of some parasitic worms are also able to attach to the fur or skin of potential hosts and become more mobile when they sense a potential host nearby. This may increase the chances of infection via larval migrans if parasites are able to sense and attach to a monkey during grooming, no matter the direction in which the grooming is being directed.

Fortunately, a number of other researchers have also recently tackled this issue. One paper in particular by MacIntosh and colleagues noted that social environment might play a role in the transmission of some helminth parasites, specifically nodular worms and threadworms, which we would not expect (MacIntosh et al., 2012). A recent study on baboons by Tung and colleagues revealed that social contact between individuals predicted the genetic makeup of individuals' gut microbes (Tung et al., 2015a). The data seem to conclusively demonstrate that social grooming behaviors have an impact on the transmission of a greater variety of parasites than previously acknowledged.

It is unclear how the parasites that Loskop vervets have affect their health. Many gastrointestinal parasites are known to be asymptomatic in many cases, with symptoms only detected in heavy infections. None of my study subjects exhibited visible signs of discomfort. I also found no correlation between detection of a parasite in a given fecal sample and whether the stool was loose, contained mucous or blood, or any other variables. The only individuals who died during the study went missing suddenly and showed no prior signs of illness. This suggests that the gastrointestinal parasites infecting Loskop vervets do not place immediate selective pressures on individual hosts, although they may degrade health and fitness over time.

Sociality – and specifically social grooming – has been selected for among primates, so clearly it must provide benefits if it also presents costs. Joan Silk presented an explanation of why sociality is selected for among primates and other mammals (Silk, 2007a, 2007b). She argued that the quantity and quality of social bonds form an individual's prospects for coalitionary support. Support from others affects an individual's dominance rank, which can influence access to food, water, and other resources that influence reproductive fitness. Berkman provided a similar explanation for the link between sociality and health in humans: social networks provide direct and indirect access to resources for improving health, influence behaviors that can improve health, and can lower levels of stress, thus improving health (Berkman, 1984; Berkman et al., 2000). It is likely that this is one mechanism that selects for high sociality among Loskop vervets.

But just as physical contact can have costs for physical health, so too can our social environments. Robert Sapolsky illustrated the various ways in which one's social environment – in both humans and nonhuman primates – can degrade health in his review of the links between health and chronic stress (Sapolsky, 2005). In addition to his own work on stress and sociality in baboons, Sapolsky presented a summary of other work indicating that primate groups that are more affiliative and social are also less stressed when measured via hormonal assays.

Fortunately, social contact can mitigate the harm from social stress by releasing pleasurable endorphins and reversing damage from stressors (for a discussion see Dunbar, 2010). It can also reduce the costs of transmission of parasites by enhancing immune function

(Marketon & Glaser, 2008). These are functions of social grooming that are often not integrated into theoretical models of the evolution of primate social behavior. When these benefits to individuals are acknowledged, the benefits of social grooming clearly outweigh the costs, except in extreme cases when highly fatal pathogens are concerned, such as the Ebola virus.

Being highly social is evolutionarily advantageous. Even with increased risk of parasite transmission, social contact like grooming reduces stress, improves health, increases access to resources, and buffers against a stressful social environment. Social grooming even comes complete with its own mitigation system, enhancing immunity and health while simultaneously promoting transmission of parasites and microbes.

12 Predation and Food Competition in Vervet Monkeys (*Chlorocebus pygerythrus*)

Lynne A. Isbell

It has generally been agreed that predation and food constitute the two most important of selection pressures on animals. As such, these ecological conditions and their effects on animals have been the focus of many empirical and theoretical studies. Predation has been held responsible for numerous adaptations, some more obvious than others. Obvious antipredator behaviors include alarm calls, vigilance behaviors, and avoidance of dangerous habitats (Seyfarth et al., 1980a, 1980b; Cheney & Seyfarth, 1981; Cowlishaw, 1997; Caro, 2005; Shelley & Blumstein, 2005; Willems & Hill, 2009). There is also some evidence that, in primates, the risk of predation is reduced for animals that are socially or locationally philopatric, have larger bodies, live in larger groups, or have better vision (Clutton-Brock & Harvey, 1977; Isbell et al., 1990; Isbell & Young, 1993; Isbell, 1994a, 2006, 2009; Isbell & Van Vuren, 1996; Olupot & Waser, 2001; Le et al., 2013, 2014). Finally, theoretical arguments posit that predation pressure was responsible for the evolution of group living, multi-male groups, and greater cognitive ability (Crook & Gartlan, 1966; Hamilton, 1971; Pulliam, 1973; Alexander, 1974; Busse, 1977; van Schaik, 1983; van Schaik & van Hooff, 1983; Terborgh & Janson, 1986; van Schaik & Hörstermann, 1994; Sterck et al., 1997; Hill & Dunbar, 1998; Zuberbühler & Jenny, 2002; Shultz et al., 2004; Zuberbühler, 2007), but conclusive evidence is difficult to obtain since there is also evidence supporting alternative selection pressures (Clutton-Brock & Harvey, 1977; Wrangham, 1980; Cheney & Wrangham, 1987).

Competition for food is considered an especially strong selection pressure for female mammals because they invest considerably more time and energy than males in their offspring (Trivers, 1972). Food competition has been linked to territoriality, intergroup competition, and the existence of female dominance hierarchies within groups (Jarman, 1974; Wrangham, 1980; Harcourt, 1987; Isbell, 1991; Mathy & Isbell, 2001; Chancellor & Isbell, 2008, 2009; Hanya, 2009). It has also been suggested as the selection pressure favoring some of the adaptations attributed to predation, including large body size, group living, variation in group size and other aspects of social organization, and greater cognitive ability (Crook & Gartlan, 1966; Jarman, 1974; Clutton-Brock & Harvey, 1977; Wrangham, 1980; Cheney & Seyfarth, 1987; Janson & van Schaik, 1988; Milton, 1988; Rodman, 1988; Chapman et al., 1995; Isbell, 2004). In fact, it may be challenging to separate the effects of predation and food competition because they can affect each other. For instance, competition for food can increase the risk of predation when it causes animals to feed in more dangerous areas (Isbell et al., 1990). Similarly, predation can increase food competition by forcing animals to use less productive areas (Cowlishaw, 1997). Predation can also reduce food competition by removing animals that consume the food (Isbell, 1994; Reznick et al., 2002).

It is against this backdrop that I have conducted most of my research on vervet monkeys (*Chlorocebus pygerythrus*). My first exposure to vervets came in 1986 when Dorothy Cheney and Robert Seyfarth offered me the opportunity to be a part of their long-term vervet project in Amboseli National Park, Kenya. Their project might have continued beyond 11 years if not for a natural change in the environment that finally came to a head during my time – fever trees (*Acacia xanthophloea*), the main tree species that vervets used for both food and shelter, were in their final stages of a massive die-off. My original research plan was to address the question of why vervets live in multi-male groups, but that changed as events unfolded. No one could have planned what I ended up studying; I was simply in the right place at the right time.

One of my first observations was that as the last of the fever trees died, larger groups outcompeted smaller groups and shifted their home ranges into the home ranges of smaller groups. Because vervets are territorial in Amboseli and do not share home ranges, the smaller groups then shifted their home ranges into areas uninhabited by other vervet groups. By supplanting the smaller groups, the larger groups benefited by gaining access to more umbrella acacias (*Acacia tortilis*), which were not preferred trees, but did provide some food and shelter and were healthy, whereas the smaller groups ended up with fewer trees, thus supporting Wrangham's (1980) theoretical argument for the effectiveness of larger groups in intergroup competition for access to foods. Another analysis suggested that the minimum effective group size in vervets was two adults of either sex regardless of the number of juveniles and infants. Groups that fell below that threshold always merged with a neighboring group (Isbell et al., 1991). However, moving into new and unfamiliar areas was costly because it was associated with a higher rate of predation. Normally, movement into unfamiliar areas is done by male vervets as they transfer from one group to another, but as entire groups shifted their home ranges or fused with neighboring groups, they also moved into unfamiliar areas, and more individuals disappeared when their groups spent more time in those areas (Isbell et al., 1990, 1993). As most of the disappearances could be attributed to suspected or confirmed predation, the results provided evidence that one benefit of philopatry – or site fidelity – is reduced predation. This was followed by another analysis that suggested that when individuals have more control over the timing of their dispersal to new areas they can lower their risk of predation (Isbell et al., 1993).

Before I arrived in Amboseli, losses of adult females and immatures to predation averaged about 11 percent per year, but during one of my two years there, it jumped to at least 45 percent (Isbell, 1990). The evidence was indirect and was based on more frequent sightings of leopards (*Panthera pardus*) by observers than in previous years, a positive association between monthly disappearances of individuals and monthly frequencies of leopard alarm calls in four of five groups, and higher rates of leopard alarm calls in those same four groups compared to previous years. Only the remains of one individual were ever found. The increase in leopard predation appeared to have been a consequence of greater leopard presence in that year, but why leopard presence had increased could not be answered. This research underscored the difficulty of studying predation when it is done by a stealthy predator, even when it occurs frequently. I concluded one of my papers with a call to study predators and primates simultaneously, "even though the relative rarity of predation events may offer small sample sizes for the effort expended" (Isbell, 1990).

Often when vervets disappeared it was when I took a short trip to Nairobi to replenish my supplies. This was not just bad luck on my part; the evidence suggested that leopards were actively avoiding me, especially earlier in the study. In the first 140 observation days, I saw a leopard only once. Then I began to see leopards more often, about once every 11 observation days. These included a male leopard and a female leopard with two cubs. The increased frequency of leopard sightings began during a 30-day period in which one vervet disappeared every 2.1 days (compared to once every 15–17 days before and after), and continued to the end of the study. On my final day in Amboseli, I was able to watch the male leopard with an unobstructed view from about 50 m away for an hour before he slowly stretched in the tree where he had been resting and sauntered away (Isbell, 1994b). His behavior suggested that he, at least, was becoming habituated to my presence by that time. However, the evidence suggests that leopards in general remained somewhat shy. When I returned from Nairobi to resume fieldwork, I saw leopards much more often in the first two days than in subsequent days, as if it took them a couple days to realize I was back to stay for a while. In addition, disappearances of vervets were negatively correlated with proximity to a ranger station with continual human presence. Leopards' avoidance of humans helps to explain why it was so difficult to witness predation by leopards (Isbell & Young, 1993).

Following my years in Amboseli, in 1992, I began what was to be a ten-year comparative project on the behavior and ecology of vervets and patas monkeys (*Erythrocebus patas*) at Segera Ranch in Laikipia, Kenya. I continued with food competition, but this time it was to explore the ecological basis of intragroup competition. It was well known that adult female vervets have stable, linear dominance hierarchies (Seyfarth, 1980; Cheney et al., 1981; Whitten, 1983), but the limited evidence from patas monkeys was contradictory, with some but not others reporting stable, linear dominance hierarchies (Kaplan & Zucker, 1980; Jacobus & Loy, 1981; Rowell & Olson, 1983; Loy & Harnois, 1988; Nakagawa, 1992; Goldman & Loy, 1997). Since these two species are sympatric in some locations and are phylogenetically closely related, they seemed ideal for a comparative socioecological study. Even better, while patas monkeys were restricted to habitat dominated by whistling thorn acacias (*Acacia drepanolobium*) away from rivers, vervets used both whistling thorn habitat and fever tree habitat along the river. If food distribution affects the nature of female competitive relationships within groups, then one might expect the agonistic behavior of vervets to become more convergent with that of patas monkeys in the same habitat. My first goal, however, was to determine whether the expression of their dominance hierarchies truly was different. After nearly four years of data collection, the results confirmed that the species do differ: female vervets at Segera, like female vervets elsewhere, had dominance hierarchies that were stable and linear, whereas female patas monkeys had dominance hierarchies that were less stable and not statistically linear. This was because patas monkeys had fewer dyads with interactions, fewer interactions per dyad, and more reversals against the hierarchy, the latter partly because maturing females tended to challenge the status quo (Isbell & Pruetz, 1998).

As predicted, the rate of agonistic interactions among vervets differed depending on the type of habitat. Their agonistic interactions were more frequent when they were in fever tree habitat than when they were in the whistling thorn habitat they shared with patas monkeys, suggesting that something about the fever tree habitat facilitated more frequent competition. At that time, since the prevailing argument was that foods that are spatially clumped foster agonistic interactions (Wrangham, 1980; van Schaik, 1989; Isbell, 1991), I expected that foods in the fever tree habitat would be more spatially clumped.

As it has not been clear what spatial scale is best for measuring food distribution relative to intragroup food competition, I quantified food distribution and abundance using a method that is organism-defined rather than investigator-defined based on the investigator's perceptions (Wiens, 1976). The movements of individuals between food items were used to describe spatial distribution, and the time spent at foods was used to describe temporal distribution on a short timescale (Isbell et al., 1998). Shorter distances between foods indicated denser foods and longer distances indicated more widely dispersed foods. Similarly, longer times at food sites indicated larger foods or longer handling times and shorter times indicated smaller foods or shorter handling times. I found that when vervets fed in fever tree woodlands, they traveled shorter total distances per unit time, moved shorter distances between food sites, and stayed longer at those food sites compared to patas monkeys in whistling thorn woodlands, conforming to expectations. In addition, when vervets fed in whistling thorn woodlands, their movements converged to some extent with those of patas monkeys in that they traveled farther per unit time and spent less time at food sites compared to when they were in fever tree woodlands. Unexpectedly, the distance vervets traveled between food sites did not change significantly between habitat types.

These results suggested that the foods of vervets in the fever tree woodlands were denser and larger or depleted more slowly, but they were not more spatially clumped than the foods in the whistling thorn woodlands. As vervets also had fewer agonistic interactions in the whistling thorn woodlands (Isbell & Pruetz, 1998), these results provided support for the concept that food drives agonistic interactions among females (Wrangham, 1980; van Schaik, 1989; Isbell, 1991), but the crucial characteristic that favors contest competition was identified as temporal rather than spatial (i.e., foods that take longer to eat are more easily usurped by higher-ranking females; Isbell & Pruetz, 1998; Isbell et al., 1998). A more conventional approach to measuring food patchiness in the same populations was consistent with these results (Pruetz & Isbell, 2000). Since then, the temporal component of food consumption has been shown to be important in facilitating agonistic interactions in several other species (rhesus macaques, *Macaca mulatta*: Mathy & Isbell, 2001; Chancellor & Isbell, 2008; ring-tailed lemurs, *Lemur catta*: Gemmill & Gould, 2008; gray-cheeked mangabeys, *Lophocebus albigena*: Chancellor & Isbell, 2009; mountain gorillas, *Gorilla beringei*: Wright & Robbins, 2014; but see ursine colobus, *Colobus vellerosus*: Saj & Sicotte, 2007).

Our organism-defined methodology for quantifying food distribution also served as a test of the hypothesis that variation in predation risk causes variation in spatial cohesion within groups and among species (van Schaik, 1989; Sterck et al., 1997). We found that although vervets were at a lower risk of predation than patas monkeys, they were more cohesive. In both species, interindividual distances mapped closely onto distances between food sites, suggesting that spatial cohesion reflects food distribution instead (Isbell & Enstam, 2002; Isbell et al., 2009).

As fate would have it, the Segera vervet study population also experienced a brief but intense period of predation, and indirect evidence pointed again to leopards as the main culprits. In less than a week, nearly half of the members of one study group of 21 individuals died of suspected or confirmed predation. Remains were found of two individuals, and our sightings of leopards and their tracks increased around that time (Jaffe & Isbell, 2010). On another occasion, claw marks on a sleeping tree, vervet body parts in the tree and in the river below, and three missing vervets in the morning provided more tangible evidence of the damage that can be done in short order to vervets (Isbell et al., 2009). It is difficult to conceive that one leopard could kill so many individuals in one night. As in Amboseli, we had sightings of a female leopard with two cubs at around the time of intense predation. Vervets may be most vulnerable when female leopards hunt with their offspring.

Both Amboseli and Segera suggest that life is challenging for wild vervets. For the Amboseli vervets, a drastic decline in their food supply followed by heavy predation reduced the study population from six groups to three by the time I had finished my study. For the Segera vervets, agonistic interactions were not as frequent either within or between groups (Cheney, 1981; Isbell et al., 1991; Isbell & Pruetz, 1998; Jaffe & Isbell, 2010), and the interpretation might be that acquiring food was less problematic than in Amboseli. However, only 3 of the 19 females born during the study survived to reproduce by the end of the study, and age at first reproduction was later in life for two of them than for Amboseli vervets living in poorer-quality home ranges (Isbell et al., 2009), suggesting food scarcity of some sort. Moreover, the fever trees along the river had begun to thin out (personal observation). Otherwise, mean annual infant mortality of Segera vervets over the ten-year study, at 48.3 percent, was typical of mammals in general (Caughley, 1966, 1977; Richard, 1985, p. 283). Many of the infants (32.1 percent) died of suspected or confirmed predation, but there were also other ways to die, including being orphaned before six months and being born in a wetter month, which may have facilitated respiratory or other diseases (Isbell et al., 2009).

After the Segera project ended, I returned to my original research question of why vervets live in multi-male groups. This is a puzzle because their closest phylogenetic relatives, patas monkeys and *Cercopithecus* spp., all live in single-male groups. Moreover, because patas monkeys, like multi-male, multi-female groups of vervets and baboons, live in semiarid habitats and spend a lot of time on the ground, gross niche differences cannot explain this. In fact, none of the factors that have been suggested to affect male membership in primate groups (i.e., poor female defensibility, high predation, cooperative home range defense, cooperative defense of females, and time constraints) can explain it. My experience with vervets in two habitats where the configuration of home ranges was influenced by fever trees located mainly around either swamps or a river, plus evidence from the Amboseli vervets that unfamiliar areas are more dangerous, as well as observations of male dispersal in the Segera vervets, helped produce an alternative explanation that was referred to as the Limited Dispersal hypothesis (Isbell et al., 2002).

Home range configuration is determined by the locations of vervet feeding and sleeping trees, which are determined by the location of standing water. Segera vervets are typical of vervets in most places in eastern and southern Africa in being restricted to living largely along rivers, whereas Amboseli has no rivers and vervets lived near swamps. As a result, Segera vervet home ranges were more linear than those of Amboseli vervets. Regardless of home range configuration, when male vervets immigrate, they typically move only into adjacent groups because of the high potential costs of dispersal. Making minor moves likely reduces the risk of predation, especially if males can make forays into the neighboring home range before permanently moving (Cheney & Seyfarth, 1983; Isbell et al., 1993). Living in linear home ranges along rivers, Segera males had the choice of only two groups into which to transfer: one or the other group on either end of their home range. By contrast, Amboseli vervets had more options because they had more neighboring groups, but despite that, males tended to transfer into groups to which former groupmates transferred (Cheney & Seyfarth, 1983). Over time, limited dispersal increases genetic relatedness between immigrants and members of their new groups as male relatives transfer to the same groups, and this has implications for infanticidal behavior, which is committed most frequently in primates by immigrant males. In the Limited Dispersal hypothesis, we suggested that closer genetic relatedness of immigrant males and members of neighboring groups would reduce infanticidal behavior by immigrants, thereby reducing aggression from resident males and females toward immigrants and allowing multiple males to coexist within groups. If this hypothesis is true, then we can predict a higher rate of infanticide in populations that have low predation and greater dispersal options. Segera vervets (with the most restricted dispersal options), Amboseli vervets (with intermediate dispersal options), and vervets in Barbados (with fewer predators and greater opportunities for gene flow than either Segera or Amboseli) (Horrocks, 1984; Horrocks & Hunte, 1986) supported the hypothesis because they presented a gradient of increasing infanticide (Isbell et al., 2002). This hypothesis needs to be tested with genetic analyses.

Twenty-seven years after leopards decimated the Amboseli study population and 15 years after leopards did the same to the Segera study population, I heeded my own call and conducted a year-long field study, this time at Mpala Research Centre in Laikipia, in which leopards, vervets, and olive baboons (*Papio anubis*) were simultaneously followed remotely with Global Positioning System (GPS) units worn on collars. Our findings challenge several proposed antipredator adaptations mentioned at the beginning of this chapter, including that larger body size confers greater protection against predators and that larger groups have an advantage over smaller groups in detecting predators. Moreover, although Mpala vervets had nearly three times the spatial encounters that Mpala baboons had with leopards, their predation rates were not correspondingly higher. The two species were, however, vulnerable at different times of the diel period: baboons were more vulnerable during the night, whereas, surprisingly, vervets were more vulnerable during the day (Isbell et al., 2018). We also found that vervets' "leopard" alarm calls not only warn conspecifics (Seyfarth et al., 1980a, 1980b), but also deter leopards, causing them to move away from vervets within minutes after the monkeys begin vocalizing, at least at night (Isbell & Bidner, 2016).

To summarize and conclude, my research on vervets has focused on understanding how ecology affects behavior. Nature, good fortune, perseverance, and serendipity have allowed me to investigate how food competition and predation affect the behavior and social organization of vervets and to use vervets as a model system for testing hypotheses and generating concepts that might apply as well to other primates. I have examined the causes and consequences of intergroup competition, ecological conditions favoring intragroup competition among females, predation, dispersal, and life history, ecological and social determinants of vervet social organization, but there is still much more to do.

ACKNOWLEDGMENTS

The research described above was supported by funds from NSF, NIH, and the University of Pennsylvania to R. Seyfarth and D. Cheney, and by funds from a University of California Regents' Fellowship in Animal Behavior, Sigma Xi, Wenner-Gren Foundation, LSB Leakey Foundation, Rutgers University Research Council, the UC Davis Bridge and Faculty Research Grant Programs, the California National Primate Research Center (through NIH grant #RR00169), and NSF (SBR 93-07477, BCS 99-03949, and BCS 1266389). I thank Trudy Turner for the invitation to contribute to this volume.

13 Vervet Monkeys' Social Learning Abilities

Erica van de Waal

The main effort of modern biology focuses on evolution and its genetic heredity basis. A much smaller literature exists on the "second inheritance system" (Whiten, 2005) that allows new behaviors to be transmitted between individuals through social learning. Biologists have defined culture as "group-typical behavioural patterns, shared by community members, that rely upon socially learned and transmitted information" (Laland & Janika, 2006). The study of animal social learning and culture has seen enormous scientific progress lately, driven by multiple factors including the results of long-term field studies (Whiten & van Schaik, 2007), new approaches to captive and experimental research (Whiten, 2009), theoretical modeling (Richerson & Boyd, 2005), and solid experimental evidence that animals can learn socially (Galef, 2012). A variety of social learning mechanisms have been identified, including stimulus enhancement (a model draws attention to an object while the individual has to learn individually its use), local enhancement (a model draws attention to a location while the subject has to learn individually what to do there), various forms of imitation (reproducing the actions of a model), and emulation (learning about the result or goal of a model's action then using one's own methods to achieve it) (Hoppitt & Laland, 2008). Such social learning provides the mechanisms underlying the cultural transmission of locally adaptive traditions (Whiten, 2005).

However, most of the research on social learning is restricted to studies with captive populations, which most easily afford the experimental controls involved. To know if the phenomena so identified are important in the wild, comparable field experiments are vital, yet they remain relatively rare (Whiten & Mesoudi, 2008; Thornton & Clutton-Brock, 2011). My contribution to the topic has been to test social learning mechanisms for the first time experimentally on a wild primate, the vervet monkey (*Chlorocebus pygerythrus*) of South Africa. With my field experiments, I have identified different social learning mechanisms and gained important insights into the rules individuals use to decide when and from whom to learn socially (van de Waal et al., 2010, 2012, 2013a, 2014, 2015; van de Waal & Bshary, 2011).

I have investigated the question of social learning now for as long as a decade, focusing on vervet monkeys. Vervet monkeys are an ideal species for field experiments on this topic as they are not endangered (allowing field experiments involving captures and feeding), are considered as opportunistic (they engage readily in experiments as soon as they know and like the food involved), and have an appropriate group structure. Vervet group structure is a multi-female/multi-male organization, with females being the philopatric sex and males dispersing first when they reach sexual maturity (around four years of age), then multiple times throughout their lives. Group size varies from a few individuals up to more than 50. The social structure is called "matrilineal," as females are closely related and stay all their lives in their natal group (except in the rare cases of group fission, when a group gets too large for its resources), where they form close bonds with other group members. This social structure is ideal for investigating the transmission of socially learned traits between a wide range of individuals both related and not.

The findings of my research extended our knowledge of social learning in wild primates primarily concerning two broad topics: (1) social learning mechanisms and (2) social learning biases (or "strategies": when and from whom to learn; Laland, 2004). My first experiment designed to test social learning mechanisms in wild vervets utilized puzzle boxes called "artificial fruits" that incorporated different options to open them in order to gain a reward (Whiten et al., 1996). These "two-action boxes" had two different openings on two differently colored ends of a box, one slide door and one lift (Figure 13.1a). The methodology is straightforward: the group

FIGURE 13.1 (a) Vervet monkey manipulating the pull door (lighter shade) and another vervet manipulating the slide door (black). (b) Vervet monkey interacting with the two-step task. (c) The three different techniques to open the food container. (d) The "vervetable": door being lifted and door being slid to the right.

member who monopolizes the apparatus is trained as a model; in this experiment, it was either the dominant female or the dominant male depending on the group. During the training phase, one of the doors is locked so that the model always demonstrates the same opening technique. After 25 successful openings by the model, I offered up to four artificial fruits to the group, this time with both doors unlocked, and the side that each group member approached and manipulated was coded. This experiment revealed that wild vervets copy manipulating the side of the box they had witnessed, thus demonstrating "stimulus and/or local enhancement" in wild primates (van de Waal et al., 2010; van de Waal & Bshary, 2010). The study also yielded insights into social learning strategies, with more attention paid to dominant females than males (van de Waal et al., 2010), and flexible manipulation abilities being found to depend on experience with human structures (van de Waal & Bshary, 2010). In observations of spontaneous feeding and grooming bouts, we observed that the bias of attention of group members toward adult females is also consistent in these contexts (Renevey et al., 2013).

My second experiment involved a more complex type of artificial fruit with two steps involved in retrieving the food reward: removing a lock on the top of the box released a cord, allowing the door to be lifted (Figure 13.1b) (van de Waal & Bshary, 2011). The results of this project confirmed the "local enhancement" mechanism described in our first experiments with group members that could observe a model manipulating the lock, whereas individuals with no prior knowledge of the task (in control groups) manipulated only the door. However, we did not find that they had learned the order of this two-step task (in contrast to chimpanzees: Whiten, 1998), as they did not start manipulating the lock directly.

Thirdly, we conducted a food-cleaning experiment inspired by the famous putative tradition of sweet potato washing in Japanese macaques (Kawai, 1965). We offered sandy grapes to six groups of wild vervet monkeys and observed whether they would clean the grapes, and if so, how they would handle them. We identified four different handling techniques for cleaning the grapes, and also that some individuals just ate them with the sand. This experiment showed that matrilines rather than entire groups were identified as the key units for social transmission of cleaning techniques (van de Waal et al., 2012). These results reflect the importance of the matrilineal (mother and offspring) substructure within a vervet group. Further detailed video analyses of this experiment revealed that naive infants copy the particular handling technique of their mother, even when foraging with other group members around (van de Waal et al., 2014). This bias of attention of young infants toward their mothers was previously suggested by observational studies correlating mother and offspring data, so here we tested experimentally for the first time the importance of the mother as the first "role model" for infant primates.

Some of our results indicated important differences between wild vervets and those in contact with humans (van de Waal & Bshary, 2010); hence, I am currently comparing their social learning skills in captivity. To test social learning mechanisms in more detail, we recently conducted more diffusion experiments in group-living vervets in sanctuaries (van de Waal & Whiten, 2012; van de Waal et al., 2013b). One captive experiment used a food container that could be opened in different ways (van de Waal & Whiten, 2012). We observed spontaneous emergence of variance in the opening techniques in four groups of sanctuary vervets: opening the lid with the mouth or with the hands or opening by pulling the cords attached at both ends of the apparatus (Figure 13.1c). Copying the part of the body used to solve a task was demonstrated for the first time in vervet monkeys and is called "bodily imitation." We also documented the spread of these three opening techniques in the different groups. After such a successful test of this apparatus in the captive setting, we tried the same experiment with our wild population, but unfortunately, all three models used the technique of opening the lid with the hands, so we had no variance for testing whether social learning could lead to bodily imitation in wild vervets.

We also extended the successful approach we developed with our simple artificial fruit (van de Waal et al., 2010), having shown that vervets can perform sliding and pulling movements. In our new puzzle box ("vervetable"), the models always acted on the same point as only one door existed, although there were three possible solutions: sliding a door sideways either to the right or to the left or opening a hatch set into the sliding door (Figure 13.1d). We ran this experiment first with four groups from sanctuaries and we found that the captive monkeys consistently maintained the majority method (even in terms of to which side to slide the door), indicating a level of social learning discrimination for monkeys that is much closer to true imitation than widely suspected (van de Waal et al., 2013b). Recently, we performed this puzzle box "vervetable" experiment with wild vervets. Following our findings that female models attract more attention than males (van de Waal et al., 2010), I trained one adult female each in three groups to open the box, with each one trained on one of the three alternative techniques. The models demonstrated 50 openings on one box using the same technique (the other options were blocked). Once the demonstration phase was completed, we offered multiple boxes to the group (this time the doors were unlocked so three different opening techniques were possible). We found copying of the method used by the model (slide versus pull) during the naive monkeys' first opening as well as over all experimental trials, which this time cannot be explained by "local enhancement," as the same knob of the same door was used by all models, so the learning mechanism involved is more complex: either imitation or emulation (van de Waal et al., 2015).

Recently, we have also been developing a totally new methodology for studying social learning in the field, using a foraging experiment where whole groups were trained to prefer one of two alternative food sources, and only a few naive individuals (infants of the year and immigrant males) were then confronted with different local norms (van de Waal et al., 2013a). In winter 2011, we provisioned wild vervet monkeys with two bowls of maize corn dyed in different colors (pink or blue), with one set initially made strongly distasteful using bitter mountain aloe. Over four sessions, 109 monkeys in four groups learned to prefer one color and to avoid the other (Figure 13.2). Months later, we offered both colored foods untreated just when a new cohort of infants was ready to take their first solid foods. Vertical social learning was strikingly demonstrated, with all infants taking only the colored alternative eaten by their mothers. Just one infant ate the formerly distasteful colored food after its low-ranking mother did. Most excitingly, ten males migrated from groups where one color was favored to a group where the opposite color was eaten, and all but one conformed directly to the new group's color preference, in marked contrast to their natal preference. The one exception was a male who rapidly achieved high rank on immigration to his new group. Possibly a different learning strategy applies in this context.

This experimental setup has been tested first over five trials and then after a six-month gap over the last two years, and is still being pursued now in order to test whether the food preference stays stable over time. The aims are not only to test the long-term stability

FIGURE 13.2 Groups trained to prefer different colors of dyed corn.

of preferences in different age/sex classes, but also to obtain more data on the potential link between the rank of immigrated males and conformity, as well as the transmission from current juvenile females to their own offspring. Furthermore, once males trained on this food preference experiment migrate to one of the two study groups untrained on this setup, I will start the experiment in these groups and perform them regularly in order to test whether the knowledgeable new male spreads his color preference in a new group. This latter situation would be very interesting to test, as until now we had entire groups trained on the task and a few naive individuals learning socially, whereas in this new context we would have a single knowledgeable individual entering a whole naive group. The preliminary results are similar in all four groups, with females being totally conservative (high-ranked females never even trying the other color, low-ranked females switching to the preferred color as soon as not outranked), whereas juveniles and males explore more of both food resources. This finding is really interesting for social learning, because if the philopatric sex (females) is so conservative, this could explain how arbitrary traditions could be maintained in neighboring groups within the same population. This experiment yielded very exciting results, with infants and immigrant males copying the local norm, and these data are much more powerful than those of the previous experiments with only one trained model and an otherwise naive group.

To conclude, vervet monkeys have proven to be an ideal species for conducting social learning experiments in the wild. We discovered that they may copy in some detail using mechanisms from stimulus enhancement to imitation and/or emulation. Our research also revealed that vervet monkeys pay selective attention toward female models and that young infants learn selectively from their mothers. We also discovered the great adaptive capacity of adult males who conform to new group norms after migration. Taken together, these findings about social learning in nonhuman primates shed important light on the emergence, propagation, and maintenance of culture, once thought to be a uniquely human trait.

Part V Life History

14 Life History of Savanna Monkeys

Trudy R. Turner, Christopher A. Schmitt,
and Jennifer Danzy Cramer

The vervets at Naivasha were bigger than the animals at any other Kenyan site. That was expected: Naivasha was the wettest, highest-altitude site and had the lowest nighttime temperatures. According to Bergmann's and Allen's Rules, respectively, the animals at the site should have been the heaviest and have had the shortest appendages. The animals at the rest of the sites *should* have weighed less and have had longer limbs, since the altitude was lower and the local ecology was hotter and drier. However, the vervets at Mosiro, a relatively high-altitude site, were smaller in some characteristics than those at Kimana, a site that was both lower in altitude and drier. Added to normal climatic factors, two of the sites were cultivated, one by local farmers and the other commercially. Ultimately, we realized that the Naivasha vervets were simply *fatter* than the others, not adapted to higher altitudes or cooler temperatures. Naivasha was the most cultivated site and the animals had access to the richest food source – it was not Bergmann's or Allen's Rules, but the local diet and energy expenditure that had the largest influence on the phenotype.

LIFE HISTORY AND MORPHOLOGY

Size matters. It is the starting point for all life history discussions. When compared across species, body size predicts the length of the interbirth interval, timing of sexual maturation, as well as age at first reproduction and entry into the reproductive community – important factors that affect fitness. In sexually dimorphic taxa like savanna monkeys, size may be a major factor that differentiates males from females. Within relatively defined parameters, size is species-specific and partially coded for by genetic factors. As such, it is subject to the

action of natural selection. What are the patterns by which organisms reach adult size? What are the forces responsible for differences in body size between individuals in a group, between populations, or between species? What do differences in adult body size mean for an individual?

Most of the work on life history has been comparative between taxa. Comparisons across mammalian taxa show that a wide range of life history traits – including neonatal weight, number of off-spring, gestation length, weaning age, maturation age, and life span – are highly correlated with body mass (r = 0.7–0.9; Read & Harvey, 1989). These scaling patterns can be seen among species and across orders, and they also scale with allometry, which is the study of the consequences of size on physical structures and biological processes, which may not vary proportionately (isometrically) with size. Allometry and scaling allow us to understand the relationships between metabolic costs, heat loss, energy costs, and diet (Kleiber, 1961; Genoud, 2002) and to detail how these differences have been widely used to explain broad, size-related differences in diet among the primates (Gaulin & Konner, 1977; Gaulin, 1978; Clutton-Brock & Harvey, 1977).

Allometric rules also govern the relationship between body size and life history variables such as birth, litter, or weaning size, which are largely considered to be controlled or linked by energetic constraints. While selection may act primarily on body size, body size, however, cannot explain all of the variation seen in life history variables. When a life history variable is plotted against body mass, some taxa still fall well above or below the regression line (Read & Harvey, 1989). These departures from the general relationship show that factors other than size guide the evolution of life history and imply further adaptive shifts. For example, primates are distinguished from other mammals by their relatively prolonged development and life span. For its body size, a chimpanzee's life span is twice as long as a cheetah's and three times as long as an impala's, and the discrepancies between the maturation periods are even more pronounced.

When size is controlled for, life history variables appear to be correlated with one another in clear patterns called "invariants" (Charnov, 1993), although the reliability of these classic invariants has been contested (Nee et al., 2005). Deviations from these patterns suggest that departures from allometry can be explained by ecological or social factors or trade-offs among life history variables (Promislow & Harvey, 1990; Kappeler & Pereira, 2003).

A negative association between two aspects of life history is an example of a trade-off, a concept that has been a central part of theories of life history evolution. A basic proposition of life history theory is that reproduction is costly (Williams, 1966). The principle of allocation proposes that limited resources result in trade-offs between investment in one versus another aspect of life history (Stearns, 1992). However, this assumption is not always upheld (Spitze, 1991); positive correlations between life history traits such as growth rate and fecundity may occur when there is phenotypic variation in resource acquisition or when resource availability is high. A high-acquisition phenotype, however, may be costly itself, a relationship that may be evident only when comparing populations differing in resource abundance (Reznick et al., 2000).

Life history theory links the events and processes of an individual's life history to demographic and evolutionary processes (Gage, 1998). This perspective emphasizes the active role of the individual in negotiating environmental challenges during development and over the life span. Trade-offs in which limited time and resources are allocated variably to growth, maintenance, and reproduction highlight the interconnectedness of traits and life events. Three of the most discussed life history trade-offs are between (1) body size and maturation (Stearns, 1992), (2) current and future reproduction (Williams, 1966), and (3) number and quality of offspring (Lack, 1947; Smith & Fretwell, 1974). A high reproductive rate is generally associated with reduced longevity (Eisenberg, 1981), and manipulations of reproductive effort alter survival (Bell & Koufopanou, 1986). Traits that confer reproductive advantages early in life are favored by selection

even if they have harmful effects later in life (e.g., antagonistic pleiotropy; Rose, 1985). Prolonged development may increase body size
and/or experience at the onset of reproduction (e.g., Janson & van
Schaik, 1993), but may also shorten reproductive life span. Large
litters or short periods of parental investment also will increase total
reproductive output while reducing offspring survival or size without
some offsetting factor (e.g., infant parking in strepsirrhines; Tecot
et al., 2012).

While life history theory has been most often used to explain
differences between taxa, there can be clear size differences between
members of a single taxon. What can we learn from comparisons
between members of a single taxon? How does one study life history
variables in populations rather than comparatively across species?
Data for specific parameters – gestation length, size, age at first reproduction – are available for most species; however, little information is
usually available for other parameters concerning the rate of growth
and development. Those data that do exist are often derived from a
very limited sample of individuals, often in captivity or for a limited
period in a single wild group, and do not reflect the range of variation
that may be present in the wild across potentially large geographic
ranges. Data from a large number of animals living in diverse habitats
within a taxon's range or over long time periods with concomitant
ecological change are extremely rare, although they are increasing
(e.g., Bronikowski et al., 2002; Altmann & Alberts, 2003, 2005;
Blomquist, 2009; Hawkes et al., 2009; Strier et al., 2010; Blomquist
& Turnquist, 2011). Most of these studies of life history variables
in local populations have been the results of years and decades of
observation. The major benefits of larger body size in a population
appear to depend on sex: in females, larger body size can confer an
increase in fecundity, while larger males appear to more readily
accrue mating benefits (Blomquist & Turnquist, 2011), and larger
females may produce better-quality offspring (Clutton-Brock et al.,
1982). For males, larger body size provides an advantage in mating
competition or female choice (Andersson, 1994, Roff, 2002). There

are, however, also a number of potential costs to having a larger body size. Achieving a larger body size typically requires either an extension of the period of growth or a more rapid rate of growth. Any delay in maturation will tend to reduce net reproductive rate, whereas the increased foraging and greater nutritional requirements needed to support such an increased growth rate may result in either higher predation or an increased vulnerability to starvation (Janson & van Schaik, 1993; Blanckenhorn, 2000). Furthermore, the availability of appropriate habitat to satisfy these requirements may, in turn, limit the maximum achievable body size (Burness et al., 2001). On the other hand, age-specific mortalities are negatively correlated with body mass in mammals (Promislow & Harvey, 1990), presumably because larger body size may limit vulnerability to smaller predators and larger individuals may be more likely to survive periods of food scarcity (Roff, 2002). Overall, there is more evidence for benefits than for costs of larger body size (Blanckenhorn, 2000).

Within species, differences in life history may reflect genetic divergence or phenotypic plasticity. Phenotypic plasticity in life history traits or environmentally based changes in life history phenotypes can evolve as adaptive responses to predictable environmental change (McNamara & Houston, 1996). Comparative studies of life history provide evidence for considerable plasticity of life history traits in primates (Lee & Kappeler, 2003). A major source of intraspecific variation in life span, for example, is variation in age at maturity. Adaptive variation in growth rates alters life history trade-offs between body size at maturity and the duration of development, since accelerated growth allows both a large body size and rapid maturation. Nutrient stress may result in slower growth, which may in turn be associated either with smaller adult size or lengthening of development to maintain a constant body size. Dominance rank and social stress are additional avenues for plasticity in reproductive maturation, although there is limited empirical evidence for such a link in primates.

Long-term studies of populations of primates are crucial to understanding life history variation, but so are studies of animals living across a species range in a variety of environments. Investigations of within-species variability can help us to better understand the normal range of development, its regulation, and its evolution. Individual variability is, after all, the source of the genetic changes that ultimately give rise to species differences (Holmes & Sherry, 1997). Interestingly, some of the best data for within-taxa comparisons come from humans (e.g., see Allal et al., 2004; Vitzthum, 2008; Stearns et al., 2010). Because of the breadth of our sample, we can examine size and other morphological, hormonal, and genetic variables that may have an influence on life history variables across low-level taxonomic categories and environmental differences.

STUDIES OF GROWTH AND DEVELOPMENT IN SAVANNA MONKEYS

Hormonal Variation

Hormones may be key to understanding the underlying proximate physiology of life history trade-offs. As pleiotropic factors that help regulate reproduction, behavior, and development, hormones have the potential to mold life history, alternative strategies, and social structure. They provide a window onto the regulation of growth and development. When physiological systems are coupled, they may participate in trade-offs that reflect life history constraints. In particular, the reproductive hormones have been thought to be the principle "architects of life history evolution" (Finch & Rose, 1995). Together with the growth hormone axis, the reproductive hormones regulate skeletal growth and sexual maturation, and their absence or adverse effects may be responsible for senescence in later life. Examining how these hormones vary with growth profiles can help identify the mechanisms regulating life history trade-offs. Hormones have deleterious as well as beneficial effects, providing a physiological basis for the tactical trade-offs posited by evolutionary biologists. Delineating the proximate mechanisms can advance our understanding of the evolution of the life history of individuals within species. These insights can also aid in the interpretation of the patterning of life history at higher taxonomic levels.

In a series of papers, Whitten and Turner (2004, 2008, 2009) examined hormonal variation in estradiol, progesterone radio-immunoassay, testosterone (T), leptin, dehydroepiandrosterone (DHEA), and DHEA sulfate (DHEAS) in Kenyan vervet populations. There is considerable evidence that T mediates trade-offs in males between current reproduction and survival-promoting mechanisms like somatic repair or immunocompetence. Gonadotropins and glucocorticoids may mediate stress-related trade-offs between off-spring size and number in female vertebrates. The hypothalamic–pituitary–gonadal axis and the growth hormone axis help to regulate pubertal development. The adrenal androgens DHEA and DHEAS are attractive candidates for the regulation of aging in primates. Human serum concentrations begin to rise in the prepubertal period, peak in early adulthood, and decline with increasing age in both cross-sectional and longitudinal samples. Individual differences are stable over time and are associated with differences in life expectancy and age-related dysfunction. Caloric restriction that extends the life span also retards the age-specific decline in DHEAS. Prepubertal levels are correlated with height, weight, and insulin-like growth factor-I (IGF-I) concentrations and have proven useful in identifying matur-ational differences in adolescents. Moreover, their more rapid decline in baboons and macaques compared to humans suggests that these hormones may be biomarkers of primate aging. The combination of information on hormonal variation with morphology, demography, and some behavior observations has provided a powerful tool for the examination of intraspecific differences in animals facing different environmental challenges.

Testosterone

A question of continuing interest in socioecology and life history is why males would share residence or reproductive access with other males in a multi-male, multi-female group. Androgens may be important to understanding the development of this social pattern since they influence energetics, body size and growth, maturation, and aging, as well as aggressive behavior. Wingfield et al. (1990) have proposed the "Challenge Hypothesis" to suggest that T is socially mediated in order to regulate aggression during the mating season, and Wingfield has demonstrated its utility in birds and other vertebrates (Wingfield, 1990, 1994). Other models, such as the limited control and the concession models, offer alternatives to the challenge model to explain the endocrine basis of primate group formation (Whitten & Turner, 2004). In the limited control model, the number of females limits a male's ability to monopolize fertile females, selecting for low basal T to avoid overt competition. Concession models rely on the observation that female reproductive interests in favoring multiple males to reduce the risk of infanticide impact the structure of the primate group. In this case, T may be positively correlated with the number of females with unweaned offspring. A high T response to infanticide attempts would be predicted. Whitten and Turner (2004) tested predictions from these models using the Kenyan vervet sample and used a novel method to simulate a rapid response to stress. An index of T response to challenge was developed using the response to anesthesia, which is known for elevating T levels within 15 minutes. Since they knew the amount of time from sedation to blood draw, they were able to calculate a response to this challenge and use it as a proxy for reproductive challenge. The limited control model predicts that the number of males would increase with the number of fertile females; however, the concession model was the best predictor of serum T patterns. Serum T in the stimulated animals increased with the number of unweaned offspring and declined with the number of females without dependent offspring, reflecting the counterbalance of availability of fertile females and infanticide risk. Both the concession and limited control models suggest male tolerance of other males, and both may be operational in determining group size and pattern.

T averages in all populations of vervets in Kenya were lower than those found in captive populations. The group with the highest levels was the group that was sampled closest to the breeding season. Differences in T levels among individual males in a social group were also observed. One of the most striking observations was that some groups had multiple males, while others only had a single male. An increase in the number of females in uni-male groups did not result in an increase in the number of males, although there was a greater T response in males in uni-male groups in the stimulated samples. In multi-male groups, one of the males had a T level higher than any other male; all other males had suppressed levels of T. In the group where male rank had been established by behavioral observation, it was the second-highest-ranking male who had the highest T level. T in multi-male groups appears to be suppressed below baseline levels, which might help explain males' tolerance of each other outside of the breeding season in these multi-male groups. Interestingly, based on DHEAS concentrations, there did not seem to be any older males in uni-male groups.

Uni-male groups were found in areas of greater human activity and in areas of higher elevation and rainfall. Rainfall has been used as a proxy for habitat richness (Chapman & Balcomb, 1998). This trend is similar to the general cercopithecine pattern of larger groups in semiterrestrial habitats with continuously distributed food resources and smaller groups in small, high-quality patches in arboreal habitats. Both fossil and molecular evidence suggest that savanna monkeys and other terrestrial cercopiths moved from an arboreal habitat to the ground. The increased need for a larger group size to aid in predator detection might have been intimately linked to the suppression of T and the increase of male tolerance.

Leptin

The protein leptin is produced by adipocytes and functions to regulate food intake and energy expenditure, as well as being an integral component in a cascade of hormones that regulate pubertal timing. Its broad range of functions suggests that it may have evolved as a signal of the adequacy of energy reserves to support reproduction, pubertal onset, and immune function (Whitten & Turner, 2008). It is the mechanism for enacting optimal resource allocation rules through its action on the gonadotropin-releasing hormone (GnRH) neurons in the arcuate nucleus (Chan & Mantzoros, 2001).

Our data allowed us to test some of the predictions of the action of leptin in a unique situation. The models of leptin action often do not account for seasonality, change in habitat, or reproductive state. Since the Kenyan vervet data were collected over the course of a year in different habitats with animals in different reproductive states, we were able to address the changing environments and their actions on leptin levels. From the total sample, leptin concentrations were available for 60 females and 56 males. The gonadal steroids, T, estradiol, and progesterone were also analyzed for these animals. Estradiol in combination with palpation and the presence of cystyl aminopeptidase (Dracopoli, 1982; Turner et al., 1987) was used to determine pregnancy.

There were clear sex differences in leptin concentrations between males and females, with females having higher and more variable leptin concentrations. In males and acyclic females, leptin was not correlated with body mass, although this was not true for lactating females. Leptin did vary with reproductive state and was positively correlated with levels of serum estradiol from pregnant females. There were site differences in leptin concentration – animals living in wetter sites with more abundant resources had higher leptin levels. This is complicated by the fact that some sites were sampled in the dry season, others in the wetter season. However, one of the low-altitude, low-rainfall sites was sampled at the end of the wet season. In males and acyclic females, leptin varied by season with a difference of four- to five-fold over the course of the year. In general, the lowest levels were found during the long, dry season and the highest levels were found in the wet season.

In general, leptin levels in animals in the wild were significantly lower than those of animals in captivity. The leptin levels in the Kenyan vervet sample reflected habitat and seasonal variation despite overall low leptin levels. They also reflected reproductive state – very high levels were found in pregnant females. Dramatic increases occurred during gestation and declined over the period of lactation. This may indicate that high leptin levels in wild populations may be unique to females, especially pregnant females.

MORPHOMETRIC STUDIES OF GROWTH AND DEVELOPMENT

Growth and development studies constitute another way to get at the critical developmental parameters that can answer questions about life history. These studies provide information on development of individuals and suggest the evolutionary forces that modify this development. Growth and development studies can also document differences in size and shape among populations of animals living under different environmental conditions. Our studies on savanna monkeys have given us one of the largest databases on morphological variation in living primate populations. We have information on animals assigned to different taxonomic categories as well as living in very different ecological conditions.

Most of the initial studies of growth and development in primates were conducted on captive animals. Captive animals can provide the best-quality data for the study of growth because the animals can be sampled longitudinally over the courses of their lifetimes. An animal in captivity can be measured every week or month. Diet can be controlled, as can environmental conditions. This kind of longitudinal data can be used to accurately determine growth curves and growth spurts (e.g., Bernstein et al., 2008; Schmitt et al., 2018). These studies, in conjunction with experimentally controlled dietary and environmental changes, have demonstrated that growth is sensitive to environmental conditions. Animals in the wild, however, do not have a controlled environment: different foods are available seasonally and may vary with stochastic climatic changes, and rank can influence access to food resources, for example. How do these differences in climate, food availability, and rank translate into growth differences in animals in the wild? And do these differences have an impact on reproductive success?

Information on growth in captive or semi-captive primates can be found in several studies, the majority of which utilized only a few individuals, such as in *Macaca mulatta* (van Wegenen & Catchpole, 1956; Wilen & Naftolin, 1976; Watts & Gavan, 1982; Schwartz et al., 1988; Turnquist & Kessler, 1989; Wilson et al., 1989; Tanner et al., 1990; Schwartz & Kemnitz, 1992), *Macaca arctoides* (Faucheaux et al., 1978), *Papio anubis* (Glassman et al., 1984; Coelho & Rutenberg, 1989), *Pan troglodytes* (Watts & Gavan, 1982), *Pongo pygmaeus* (Fooden & Izor, 1983), *Cercocebus albigena* (Deputte, 1992), and *Mandrillus sphinx* (Wickings & Dixon, 1992; Setchell et al., 2001). These initial studies indicated that body size and the pattern of development of sexual dimorphism seems to be species specific (e.g., Leigh, 1992), although these data sets, by their nature, ignore the variation that might be present across the entire geographic range of each species. Sex-based differences in body size in these taxa appear to result from differences either in the rate of growth or in the duration of growth in the two sexes, known as bimaturism (Gavan & Swindler, 1966; Shea, 1986; Leigh, 1992).

Recent work in captive savanna monkeys has expanded on this literature by characterizing growth in a large, longitudinal study of animals in the Vervet Research Colony housed at Wake Forest School of Medicine, using the pedigree of the colony and whole-genome sequencing of over 700 individuals to link these patterns to regions of the genome. Schmitt et al. (2018) measured over 900 captive savanna monkeys and modeled their lifetime growth in body mass and crown-to-rump length, as well as traits relevant to clinical measures of obesity like waist circumference and body mass index (BMI), from 2000 to 2015. They found that not only were all measured adult traits – including mass and skeletal measures – highly heritable, but so too were the parameters of growth leading to those adult traits. Although a sex-specific genomic analysis was not conducted, they did find the same bimaturism leading to sexual dimorphism noted in earlier studies, while also noting that bimaturism was largely responsible for the differences between obese and nonobese individuals in the same sex. These growth patterns and adult differences in body size and mass were also shown to be the result of both genetic and environmental factors. Regions of the genome associated with human traits such as obesity, metabolic function and disorders, and somatic growth showed genome-wide suggestive linkage to these traits, while the most important environmental factor associated with growth and adult size appeared to be maternal diet while the individual was *in utero*. Offspring gestating while mothers were fed a calorically challenging diet were significantly larger in body mass and BMI as adults, suggesting that maternal programming linked to dietary variation has a significant influence on growth and adult size. This *in utero* effect, however, only appeared to influence male outcomes. Schmitt et al. (2018) suggest that these results indicate both thrifty genotype (Neel, 1962; Prentice et al., 2008) and thrifty phenotype (Hales & Barker, 2001; Metcalfe & Monaghan, 2001; Wells et al., 2007) explanations for variation in body size in vervets, noting that the presence of genomic regions associated with increased body mass and the appearance of maternal programming to increase fat deposition in lean times have both been proposed as adaptations to food scarcity. That the significant effect on body size of maternal diet while *in utero* only occurs in males may also fit into an adaptive framework, and Schmitt et al. (2018) cite the literature of sex-biased parental investment in offspring (e.g., Clutton-Brock et al., 1985; Cockburn et al., 2002) as a potential framework by which to understand this variation. Whether these mechanisms are responsible for producing the variation in body size and shape seen across widespread primate populations exposed to variation in food availability and climate remains to be investigated in the wild.

The numerous studies of growth in wild populations of sexually dimorphic primates also appear to support the pattern of bimaturism seen in captive studies, but with the exception of a few long-term studies such work relies predominantly on cross-sectional data or data from just a few field seasons; for example, baboons, including *Papio hamadryas hamadryas* (Sigg et al., 1982; Jolly & Phillip-Conroy, 2003), *P. h. anubis* (Altmann et al., 1993), *P. h. cynocephalus* (Altmann & Alberts, 2005), and *P. h. ursinus* (Johnson, 2003); macaques, including *Macaca sinica* (Cheverud et al., 1992), *M. assamensis* (Berghanel et al., 2015), *M. fascicularis* (Schillaci et al., 2007), and *M. ochreata* (Schillaci & Stallmann, 2005); gelada baboons, *Theropithecus gelada* (Lu et al., 2016); sifaka, including *Propithecus diadema* (Ravosa & Hylander, 1993), *P. edwardsi* (King et al., 2011), *P. tattersalli* (Ravosa & Hylander, 1993), and *P. verreauxi* (Ravosa & Hylander, 1993; Richard et al., 2000); gorillas (Breuer et al., 2009); black-and-white ruffed lemurs, *Varecia variegata* (Baden et al., 2008); night monkeys, *Aotus azarai* (Huck et al., 2011); and savanna monkeys on the island of St. Kitts (Horrocks, 1986) and in Kenya (Turner et al., 1997; Whitten & Turner, 2004). This is, in part, because obtaining longitudinal data for wild populations of animals is not easy or practical. A researcher or team of researchers would have to sample animals on a fixed schedule over the course of several years. With animals the size of savanna monkeys, repeated sampling could entail either multiple captures or habituation to some measuring device in the field. This could lead to a number of inter-vening complications: animals might become trap shy and not return to the traps, or repeatedly trapping animals to obtain these data might not meet with current ethical guidelines, for example. Habituating animals to a scale or some other measuring device is a possibility; however, only weight is available through this method (e.g., Uehara & Nishida, 1987). In order to obtain body size measurements, the animals must be handled, although new techniques using photogrammetry to obtain size measures from a distance are increasingly being used with nonhuman primates (e.g., Rothman et al., 2008; Kurita et al., 2012; Galbany et al., 2016; Lu et al., 2016).

When longitudinal sampling is not an option, cross-sectional data constitute an alternative method of obtaining information on growth and development. Animals are sampled at a single point in time, placed into age classes, and classes are compared. Studies on humans (e.g., Harrison et al., 1988) have shown that cross-sectional data can smooth out growth spurts because they provide an average of many individuals. Even so, cross-sectional information can begin to give a general picture of development that is useful and can reveal larger patterns of growth, although it is not as finely grained and informative as longitudinal data (German, 2004).

Like longitudinal sampling, sampling animals of the same species living in different environments is also rare. Because of work with captive groups and, more importantly, because of work with human populations living in very different environments, we know that the environment can alter patterns of growth (Bogin, 1999; Stinson, 2012). Humans are probably the best primate example of alterations of growth patterns due to environment. Since the 1960s, biologists have been studying human variation in populations living under some kind of environmental stress (e.g., high altitude, cold, and extreme heat; Frisancho & Baker, 1970; Leonard, 1989). They have also looked at populations that have been isolated from other populations for relatively long periods of time, including island populations of the South Pacific and native populations of South America (Frisancho, 1978; Ulijaszek, 1993). One of the best-known populations living under extreme environmental stress are the Quechua Indians of Peru. These individuals live at extremely high altitudes where oxygen pressure is low, temperature is low, and nutrition is poor. Some of the phenotypic differences in individuals living in this environment include slow growth and a lack of a defined growth spurt, but a barrel chest. Determining whether these differences are the result of a genetic adaptation or a developmental acclimatization has occupied researchers in human biology for 40 years (Walker et al., 2006).

With wild nonhuman primates, the ways in which environmental differences influence the pattern and rate of development and dimorphism in any particular species is not well-known, except for in long-studied populations like the Amboseli baboons (Altmann & Alberts, 2005). As could be expected given differing nutritional regimes, the rate of maturation for animals in the wild is not the same as the rate for animals in captivity. In captivity, where food is readily available, animals mature at a more rapid rate (Altmann & Alberts, 1987; Phillips-Conroy & Jolly, 1988; Eley et al., 1989). Adult body size also differs (Turner et al., 2016). Other studies of captive animals living in diverse conditions and differing dietary regimes (Faucheux et al., 1978; Schwartz et al., 1988; Schmitt et al., 2018) confirm the role of the environment in maturation and development. In the wild Amboseli baboons, rates and patterns of development have been observed to vary with rank (Altmann & Alberts, 2003) and access to human-mediated resources (Altmann & Alberts, 2005).

There have been some studies of morphological variation from animals within restricted or limited species ranges in the wild. These populations of primates may either be those with more limited geographic distributions (Japanese macaques: Uehara & Nishida, 1987; booted macaques: Schillaci & Stallmann, 2005; chimpanzees: Pusey et al., 2005; ring-tailed lemurs: Cuozzo & Sauther, 2006) or those that have wider distributions but are housed in semi-naturalistic environments (rhesus macaques: Turnquist & Kessler, 1989; Blomquist & Turnquist, 2011; mandrills: Setchell et al., 2001, 2006; mangabeys: Deputte, 1992; baboons: Coelho & Rutenberg, 1989).

A third source of information on primate morphological variation derives from a handful of comparative studies that rely on information about widely ranging species that are derived from a single locality. These include the work of Jolly and Phillips-Conroy (2003) on baboons in Awash Park, Ethiopia; Altmann and colleagues on baboons in Amboseli National Park, Kenya (Altmann et al., 1981, 1993; Charpentier et al., 2012); Strum and colleagues on baboons in the Rift Valley area of Kenya (Eley et al., 1989; Strum, 1991), Johnson (2003) in Botswana, and Sigg et al. (1982) on hamadryas baboons. Information is also available for localized populations of macaques from Singapore (Schillaci et al., 2007), Kojima Island, Japan (Hamada et al., 1986), and Polonnaruwa (Cheverud et al., 1992). Only rarely is information available from free-ranging animals living in multiple locations within the species range collected by the same researchers using comparable protocols (e.g., Turner et al., 1997, 2016). Information on intraspecific differences in species living in diverse habitats with differences in latitude and rainfall indicate that sometimes body weight or mass and appendage size follow Bergmann's and Allen's Rules, while sometimes it does not (Hamada et al., 1986). In macaques, in general, appendages did reduce in length with increased latitude, increased altitude, and decreased temperature as body size increased (Hamada et al., 1986). But in groups of *Macaca nemestrina* (Albrecht, 1980) and *M. fascicularis* (Fooden & Albrect, 1993), Bergmann's Rule did not hold, and in *M. sinica*, no ecogeographic effects were noted (Albrecht et al., 1990). In baboons, Popp (1983) noted a correlation between body size and rainfall, and Strum (1991) reported that cultivated food influenced weight (as also seen in Altmann & Alberts, 2005).

While trapping, handling, and measuring animals can provide extraordinary information, these processes are not without difficulty. These projects involve complicated logistics, permits, and ethics committee clearance. Even when they are initiated, there can be procedural difficulties. Sexing infant or juvenile animals often can only be accomplished with training and practice. Age can generally only be determined by dental eruption sequences, and identifying the teeth that are present and the state of eruption takes training. Teeth erupt over the course of a period of four to five years, some teeth cluster together, and some teeth erupt over a four- to six-month period. We ultimately chose a dental aging system that grouped animals into broad categories based on dental eruption patterns (Table 14.1), which gives a somewhat continuous temporal division between dental age categories. Information that allowed us to determine this pattern was obtained originally from Ockerse (1959), subsequently modified by Bolter and Zihlman (2003), and finally altered in accordance with our own experience and observations. Ockerse examined more than 3000 captive vervets in South Africa to define the dental eruption pattern. We were able to use that sequence to age animals into eight

Table 14.1 *Dental eruption sequence and age class.*

Age class	Teeth present, fully erupted, upper and lower
1	All deciduous
2	All deciduous, M 1
3	M 1, I 1, I 2
4	M 1, I 1, I 2, M 2
5	M 1, I 1, I 2, M 2, P 3, P 4
6	M 1, I 1, I 2, M 2, P 3, P 4, C
7	All adult teeth present
8	Teeth very worn

I = incisor; C = canine; M = molar; P = premolar.

dental age classes. One of the difficulties of using dental eruption sequences is that, once animals reach the adult age category, it is not possible to determine age until excessive tooth wear suggests very advanced age. There are thus years in the adult class when no further distinctions can be made. Dental development is often presumed to be correlated with skeletal growth and development, although this relationship is, at times, complicated (Šešelj, 2013). The use of dental age (rather than absolute age) as an independent variable against which body weight is plotted may counteract some of the distortion of the weight-gain curve that results from pooling data from early- and late-maturing individuals in a cross-sectional sample (Tanner, 1962; Harrison et al., 1988).

In Ethiopia, the only measures we obtained were weight and body length. In Kenya, we took a series of ten morphometric measures that we termed "classic." In all later collections, we obtained an additional 20 measures that were suggested by the Bones and Behavior Working Group (2015). All animals were measured by the same person or by one of two other people who were trained by that person. Training involved two people measuring many animals together until measurements were concordant. The animals were weighed on a baby scale or a hanging scale to the nearest 0.5 g. Measurements were taken with a standard measuring tape and spreading and sliding calipers. Classic measures included:

Tail – base of the tail to the tip of the tail, stretched and including the hairs at the end.

BL – body length – external occipital protuberance to the base of the tail along the curve of the body.

CG – chest girth – greatest circumference of the chest.

BB – body breadth – greatest distance between tips of longest fingers when arms were spread. This is arm span.

UA – arm – acromioclavicular joint to olecranon process of ulna when flexed.

LA – olecranon process of ulna to point of flexion at carpus.

Hand – point of flexion at carpus to tip of middle digit.

UL – thigh – highest point of greater trochanter to mid-point of disto-lateral margin of lateral condyle of femur.

LL – lower leg – midpoint of disto-lateral margin of lateral condyle of femur to tip of heel.

Foot – distance along plantar surface heel to tip of toes.

There were three processes we used in order to determine whether a female was pregnant: (1) manual palpation; (2) the presence of a specific cystyl aminopeptidase phenotype shown to be associated with pregnant females (Dracopoli & Brett, 1982); and (3) lactation in preparation for pregnancy. This final method we felt not to be as reliable as the first two, since it left open the possibility of residual lactation following infanticide or infant death. If we determined that the female was pregnant (using only methods 1 and 2), we subtracted 200 g from her weight. Two-hundred grams is somewhat less than the weight of a newborn vervet monkey. We were also able to score females as nulliparous, primiparous, or multiparous on the basis of nipple elongation (Lewis et al., 1981).

The means and standard deviations for weight and all morphometric measurements for the Kenyan animals for each age class and sex were first presented in Turner et al. (1997). They are included as part of Table 14.3, which gives weight for each age class for vervets in different parts of their range. For the Kenyan vervets, differences between means were assessed for significance by t-tests with Bonferroni's correction for multiple tests, while for comparisons between all savanna monkeys, we used a Welch analysis of variance (ANOVA) with a Games–Howell Tukey post hoc test.

THE KENYAN VERVETS: SIZE AND GROWTH

Adult males are larger and heavier than adult females, while in infancy the two sexes are the same size. When does this differentiation between males and females begin? What does this have to do with their life histories and their entry into the reproductive community?

Males and females in this population do not differ in size until they reach dental age class 4 or 15–18 months of age. Females between the ages of 18 and 24 months (age class 5) do not seem to gain weight or grow in limb proportions. They only begin to grow again, initiating the pubertal growth spurt, after 24 months, and then grow substantially until 36–40 months (age class 7) of age. By 40 months, they have reached nearly adult size. There is relatively little growth beyond that time, and this growth is seen only in females at two sites – Naivasha and Kimana – where cultivated food is available.

While female growth levels off between two and three years of age, male growth continues. In fact, males grow considerably during this time period and continue to grow until they are fully adult dentally (age class 7) at four years of age. In contrast to females, males at all sites continue to grow in all body segments and in weight. Size dimorphism then can be accounted for by bimaturism, with males continuing to grow from 24 to 48 months of age, while females do not.

The age of divergence of growth rates and the velocity of growth after the divergence can be understood through the lens of life history. Females enter the reproductive community as subadults at around three years old, earlier than do males (Cheney & Seyfarth, 1983). This is underscored by the observation that primiparous females (as defined by nipple elongation) have a dental age of approximately 40 months (Lewis et al., 1981). Given that gestation is approximately

22 weeks (Rowell & Richards, 1979), females first become pregnant at approximately 36 months. In addition, STLV-I and SIVagm antibodies first appear in females at about 36 months of age (Dracopoli et al., 1986; Ma et al., 2013). These are both transmitted sexually and occur in females at younger ages than they do in males.

Females also remain in their natal group throughout their lives, while males emigrate, first as subadults and then as adults (Cheney & Seyfarth, 1983). Cheney and Seyfarth indicate that males move to groups where male kin have migrated before them, permitting them to join kin in forming coalitions in order to successfully mate. It is thus not as easy to define male entry into the reproductive community with the precision we can define it for females, but it does seem that coalition formation is important for male mating success. Males and females reach skeletal maturity at about the time they begin to reproduce, but that time is younger for females than it is for males.

That females continue to increase in body weight and some morphometric characteristics at some sites in Kenya and not at others suggests that there exist intraspecific site-to-site differences in vervets. These intraspecific differences, especially those in female body mass, could clearly have an effect on life history variables, such as number and quality of offspring. Between sites, females show significant differences in body weight, body length, body breadth, chest girth, hand, upper arm, lower arm, upper leg, lower leg, and foot lengths – everything but tail length – and these differences do not appear until age class 7. Male differences between sites can only be seen in body length and breadth and tail length, although these are not significant and are only found in adults and very young males.

Several of the variables show significant differences not just between sites, but also between groups at a single site. In order to determine where the major part of the variation occurs (between sites, between groups, or between individuals), the sum of squares of each variable is divided by the total sum of squares. For body weight, the greatest variation is between sites, then between individuals within groups. For body breadth, upper arm, upper leg, and lower leg lengths, the greatest variation occurs among individuals within groups and then between groups at site. For chest girth, variation is greatest among individuals within groups and then between sites. When using pairwise comparisons (Bonferroni/Dunn t-test), no significant differences are found between Samburu and Mosiro for any variable. The greatest number of significant differences separate females at Samburu from those at Naivasha. Chest girth is the only variable that separates females at Naivasha from females at every other site.

Dimorphism in the adult body is greater at Samburu and Mosiro than it is at Kimana and Naivasha. Samburu is a lowland site while Mosiro is a highland site, yet both are uncultivated. Kimana is a lowland site while Naivasha is a highland site, yet both are cultivated. It seems that, in this case, the real differences in the causes of ultimate body size are not in altitude or rainfall, but in access to cultivated food crops. These differences run contrary to expectations from Bergmann's Rule, which states that "within a polytypic warm-blooded species, the body size of the sub-species usually increases with decreasing mean temperature of its habitat" (Harrison et al., 1988), and Allen's Rule, which extends Bergmann's Rule to say that "in warm-blooded species, the relative size of exposed portions of the body decreases with the decrease of mean temperature" (Harrison et al., 1988). If Bergmann's and Allen's Rules held, the animals at the highest sites would be the heaviest in weight and the largest in chest girth and would have the shortest tails and limbs. In addition, there should be a gradient from highest to lowest sites. Alternatively, one might suggest that sites where there was greater rainfall and a more abundant food supply would have animals that were heavier and larger in all body segments than at sites that were drier.

Body weight in females is the only measure that increases with increasing altitude, decreasing temperature, and increasing rainfall. The only other measure that follows Bergmann's and Allen's Rules in this population is tail length, with males at Naivasha having the shortest tails. No cline is observed in the size of any variable except tail length – as temperature drops and altitude increases, tail length decreases in both males and females. The interesting aspect is that at Naivasha, where food is abundant, the animals are not larger in all body measurements. In fact, they are only heavier. The main difference between Naivasha and all other sites is the increased reliance on cultivated food. The addition of human food to natural food may cause animals to grow faster, reach their final adult weight at an earlier age, and have a higher final weight (Sadlier, 1969; Howell, 1979; Bronson, 1989; Strum, 1991), but it does not seem to influence skeletal growth.

The closest analogy for the Naivasha vervets is from the baboons at Gilgil in Kenya and Amboseli. Some troops of baboons at Gilgil were crop raiders while others were not. In the troops that were crop raiders, the females were heavier. Strum accounts for this difference by the fact that females remain in their natal group their entire lives, having constant access to cultivated food. Altmann et al. (1993) found that at Amboseli one troop raided the garbage dump while others did not. The females of the garbage dump troop were 50 percent larger than females in other troops. The differences between males were, as initially observed, not so great (although, with time, a larger impact was observed; Altmann & Alberts, 2005).

One could ask whether the weight gain in females was due to an increase in available food or a decrease in activity. Vervets not exposed to human food eat flowers, seeds, leaves, and other small objects. Smith (1978) found in populations of *Alouatta* on the same type of diet that only about 50 percent of these calories were assimilated. Human food, on the other hand, is larger and does not have protective toxins or other physical protective features and has greater digestibility. Altmann et al. (1993) found that the caloric intake of the garbage dump troop at Amboseli was approximately equal to that of animals feeding away from the garbage dump. The difference was in the amount of physical activity required to obtain food, which dropped markedly in the garbage dump troop. Garbage feeders expended at least 16 percent less energy in obtaining food, suggesting that the increase in weight was due to reduced energy expenditure and not increased caloric intake.

The Naivasha situation is similar to that of the garbage dump group in Amboseli and the crop raiders in Gilgil: human food is freely available without considerable energy expenditure. This sex-limited access to human food due to philopatry, however, is not true of these vervets. Males in Naivasha who changed groups went to other groups around the lake where cultivation continued to take place. Females do, however, seem to be more responsive to the availability of food and the decrease of activity. It is possible that male activity may take other forms and so keep energy expenditure relatively high.

LIFE HISTORY TRADE-OFFS IN THE KENYAN VERVETS

By using both morphometric and hormonal data, it is possible to describe the pattern of life history trade-offs, particularly the trade-off between growth and reproduction, in the Kenyan vervets. In both males and females, growth was faster at the site where resources were more abundant. Age pyramids for animals at Samburu and Naivasha indicate that there are higher proportions of males in the younger age classes at both sites. This is consistent with patterns of higher male mortality throughout development. Survivorship was higher at Samburu than at Naivasha, leading to the suggestion that growth should be faster and reproductive maturation earlier at Naivasha. Across sites, there was considerable variation in the timing of male reproduction, as indicated by testicular descent and T levels. For both of these parameters, Naivasha indicated the most rapid maturity. DHEAS levels suggest a lack of older males at the Naivasha site, although verifying this with other parameters was not possible.

Estradiol, which stimulates growth at low levels but terminates growth at high levels, is implicated in the fusion threshold, an indicator of puberty. These levels differ between sites. Presumably due to earlier pregnancies, females at Samburu crossed the threshold much earlier than females at Naivasha. At some sites, females began reproducing at two years of age, but at other sites not until four years of age. For both males and females, the earliest onset of reproduction was at the site of the shortest duration of growth.

The combined picture that emerges from a comparison of data on Kenyan vervets is that growth is more rapid when more food is abundant, but this rapid growth comes at a cost – males at these sites are usually found in uni-male groups. T levels are higher and death is earlier. Females at sites where food is less abundant reproduce earlier but live in groups with many males and many females.

The information from Kenyan vervets indicated that anthropogenic factors have a substantial effect on body size and weight. But what happens if we extend the hypotheses generated by the Kenyan data to a much larger data set of animals that live much further apart from each other, inhabit more extreme environments, and belong to different taxa? The material amassed by the International Vervet Research Consortium allows us to do just that.

BROADER PATTERNS IN THE GENUS *CHLOROCEBUS*

Hundreds of additional samples were added to the morphology data set as a result of the extensive sampling of the International Vervet Research Consortium and were recently analyzed and published by Turner et al. (2018). With this study, what had already been one of the largest samples of morphological variation of wild primates more than tripled in size. Table 14.2 indicates the number of samples and the demographic distributions of the animals now included in the sample. Although the original data from Ethiopia on 139 animals were limited in the number of parameters examined, they were added to the more extensive Kenya data on 352 animals. Measurements on all 30 morphological traits from 713 animals from South Africa and 409 animals from St. Kitts were included, giving us a total of 1613 animals. Of these, 748 (288 female, 460 male) were immature animals. There was a small number of samples from Botswana, Zambia, Ghana, and The Gambia, but the numbers were not sufficient to be included in further statistical models. We used a Welch ANOVA with a Games–Howell Tukey post hoc test to examine

Table 14.2 Summary statistics for vervet morphometric data

Taxon	Location	Sex	Dental Age	n	Body Mass (kg)		Body Length (cm)		n	Thigh (cm)		Lower Leg (cm)		Foot (cm)	
					Mean	S.D.	Mean	S.D.		Mean	S.D.	Mean	S.D.	Mean	S.D.
Ch. aethiops	Ethiopia	Male	1	8	0.55	0.15	18.33	2.08	–	–	–	–	–	–	–
(n = 139)		(n = 69)	2	16	1.25	0.24	25.66	2.02	–	–	–	–	–	–	–
			3	4	1.61	0.11	30.20	2.72	–	–	–	–	–	–	–
			4	3	1.72	0.36	29.70	2.10	–	–	–	–	–	–	–
			5	4	2.33	0.11	32.23	2.80	–	–	–	–	–	–	–
			6	5	3.54	1.06	36.32	2.48	–	–	–	–	–	–	–
			Adult	29	4.02	0.68	38.18	2.91	–	–	–	–	–	–	–
		Female	1	2	0.24	0.13	15.45	0.64	–	–	–	–	–	–	–
		(n = 106)	2	12	1.29	0.13	25.72	2.27	–	–	–	–	–	–	–
			3	4	1.79	0.42	28.13	2.78	–	–	–	–	–	–	–
			4	4	1.84	0.42	30.15	3.99	–	–	–	–	–	–	–
			5	–	–	–	–	–	–	–	–	–	–	–	–
			6	7	2.16	0.19	31.43	0.93	–	–	–	–	–	–	–
			Adult	41	2.70	0.51	33.28	2.02	–	–	–	–	–	–	–
Ch. pygerythrus	Kenya	Male	1	31	0.84	0.31	21.84	5.06	32	8.95	1.47	9.20	1.57	9.05	1.22
(n = 1065)	(n = 352)	(n = 183)	2	37	1.38	0.21	28.34	2.24	37	10.74	1.07	11.03	1.19	10.38	1.20
			3	20	1.66	0.18	30.33	2.47	20	11.58	0.92	12.13	0.90	10.93	0.69
			4	22	2.01	0.29	32.64	2.11	22	12.39	1.03	12.91	0.93	11.78	1.23
			5	6	2.46	0.42	34.33	2.34	6	14.00	0.84	14.33	0.82	13.25	0.52

		6	9	3.10	0.66	37.33	1.48	9	15.67	0.79	16.11	1.05	13.39	0.99
		Adult	58	4.34	0.64	41.60	2.98	58	16.32	1.18	16.56	0.99	14.04	0.97
	Female	1	6	0.92	0.10	23.42	1.53	6	9.58	0.97	10.05	0.91	9.13	0.67
	(n = 169)	2	16	1.40	0.32	28.33	2.63	16	10.81	0.98	11.38	1.18	10.47	0.85
		3	9	1.63	0.11	30.61	1.96	9	11.50	0.71	12.17	0.79	11.33	0.35
		4	12	1.73	0.25	31.67	1.89	12	11.79	1.10	12.38	1.09	11.17	0.72
		5	2	2.38	0.53	35.00	0.00	2	15.00	2.12	14.50	0.71	13.25	0.35
		6	31	2.41	0.33	34.47	2.28	31	13.65	0.97	13.92	0.98	12.05	0.53
		Adult	93	2.97	0.48	36.77	2.91	94	14.07	1.14	14.23	1.08	12.14	0.86
South Africa	Male	1	27	1.37	0.42	24.46	3.52	28	11.09	1.37	12.21	1.60	10.46	0.90
(n = 713)	(n = 343)	2	78	1.86	0.45	28.51	3.07	79	12.50	1.19	13.61	1.23	11.39	1.07
		3	39	2.24	0.48	30.09	3.63	39	13.72	1.16	15.05	1.17	12.17	1.01
		4	43	2.74	0.55	33.85	3.72	43	14.71	1.21	16.21	1.16	13.13	0.98
		5	21	3.15	0.53	34.71	2.94	22	15.48	1.46	17.30	1.31	13.84	0.92
		6	31	4.43	1.17	38.98	3.84	31	17.32	1.34	18.84	1.19	14.71	1.22
		Adult	104	5.69	0.73	40.81	3.56	104	18.50	1.24	19.95	1.08	14.92	0.80
	Female	1	27	1.27	0.33	23.81	2.41	26	10.92	1.33	12.10	1.32	10.26	0.70
	(n = 370)	2	50	1.74	0.43	27.87	3.07	50	12.18	1.18	13.40	1.26	11.22	0.98
		3	30	2.03	0.39	28.65	3.59	30	13.25	0.94	14.42	0.86	11.53	0.72
		4	37	2.50	0.51	31.97	3.04	37	14.04	1.42	15.01	1.23	12.30	0.89
		5	8	2.79	0.62	33.50	3.37	8	15.13	1.30	16.31	1.33	12.94	0.98
		6	53	3.45	0.56	35.76	2.67	53	15.42	0.89	16.65	0.67	13.02	0.72
		Adult	165	4.09	0.66	37.18	3.30	166	16.00	0.97	17.06	0.92	13.12	0.85

(*continued*)

Table 14.2 (*cont.*)

Taxon	Location	Sex	Dental Age	n	Body Mass (kg)		Body Length (cm)		n	Thigh (cm)		Lower Leg (cm)		Foot (cm)	
					Mean	S.D.	Mean	S.D.		Mean	S.D.	Mean	S.D.	Mean	S.D.
Ch. sabaeus	St. Kitts &	Male	1	21	0.96	0.21	22.67	5.90	22	9.50	0.83	10.93	1.82	9.59	0.81
(n = 409)	Nevis	(n = 212)	2	28	1.37	0.45	26.14	3.95	28	11.20	1.30	12.37	2.10	10.70	1.05
			3	19	2.05	0.39	29.66	2.11	19	13.11	1.17	14.53	1.18	11.84	0.80
			4	15	2.36	0.46	31.40	2.85	14	13.89	1.38	15.54	1.41	12.54	1.18
			5	16	2.69	0.46	32.91	3.29	16	14.59	1.05	16.00	1.06	13.25	1.83
			6	37	4.04	0.92	36.77	3.84	38	16.84	1.47	18.58	1.39	14.45	0.85
			Adult	90	5.67	0.74	40.41	3.08	90	18.24	0.83	19.81	0.76	15.11	0.67
		Female	1	17	0.89	0.23	22.26	1.94	18	9.42	0.75	10.61	0.99	9.33	0.75
		(n = 183)	2	16	1.62	0.28	26.94	2.29	17	11.62	0.82	12.79	1.06	10.68	0.79
			3	9	2.28	0.49	31.78	3.36	10	13.10	1.82	14.60	1.87	12.05	1.23
			4	14	1.99	0.51	30.07	2.37	13	13.23	1.24	14.43	1.22	11.86	0.77
			5	10	2.37	0.20	31.70	2.20	10	14.00	0.78	15.56	0.73	12.20	0.59
			6	53	3.41	0.68	34.59	2.99	53	15.31	0.80	16.45	0.79	13.08	0.97
			Adult	64	3.54	0.81	34.67	2.57	64	15.33	0.92	16.47	0.75	12.91	0.79

differences in mean mass and limb length across age/sex classes and location, while loess curves were used to model pre-pubertal growth and linear mixed models or beta regression mixed models were used to analyze body mass and relative limb segment lengths, respectively (Turner et al., 2018).

The earlier close analysis of the Kenyan vervets discussed above indicated that form was modulated in part by local environmental forces as well as human impacts (Turner et al., 1997); the inclusion of animals separated by not only larger geographic distances, but also from different taxa led to similar results with an even stronger signal from climate and local ecology. In this analysis, each of a smaller subset of traits was analyzed separately. Those that indicated overall body condition, such as mass, were considered separately from those that were primarily size related, such as upper and lower leg length and foot length relative to body length.

Stronger than the evidence seen in just Kenya, worldwide variation in savanna monkey mass and limb proportions appeared to follow both Bergmann's and Allen's Rules, although the actual size differences in limb proportions were very small (Turner et al., 2018). Females appeared to follow these patterns more strictly than males, for whom sexual dimorphism appeared to disrupt the ecogeographic and climatic patterns in body size hypothesized by Bergmann's Rule, most notably in St. Kitts and Nevis. In analyses only using geographic variables for Bergmann's and Allen's Rules, there was also a significant effect of human impacts on body size, with animals experiencing greater human impacts being significantly heavier than those in areas relatively isolated from human contact. In analyses incorporating climatic variables, however, geographic and human impact variables did not have as significant an impact as certain climatic variables – although these variables were inconsistent across traits and provided inconsistent support for Bergmann's and Allen's Rules – like the minimum temperature of the coldest month (negative relationship with body mass in males; negative relationship with limb length in both sexes), precipitation seasonality (positive relationship with body mass in males; negative relationship with limb length in males), mean annual temperature (negative relationship with body mass in males; positive relationship with limb length in both sexes), and annual precipitation (negative relationship with limb length in males), to name just a few. These inconsistent relationships with climatic variables and loss of support for geographic variables when climatic variables were included suggest that local ecogeographic context may have unique effects on these traits that cannot be captured by global models; localized effects may be too context-specific or idiosyncratic for the models used in this paper to capture, despite the considerable sample size. As in Kenya, closer investigations of local context in order to better understand population pressures are more likely to be readily interpretable.

Sexual dimorphism in body mass occurred in all populations; however, the degree of sexual dimorphism was different in each. The greatest sexual dimorphism was found in *Ch. sabaeus* from St. Kitts and Nevis, while the least pronounced sexual dimorphism was in *Ch. pygerythrus* from South Africa. This is particularly interesting since Rensch's Rule (Rensch, 1950; Clutton-Brock et al., 1977) suggests that taxa with the largest average body mass should also show the largest dimorphism (although there have been notable exceptions in primates; Gordon, 2006), and the South African animals are the heaviest overall. Except in *Ch. aethiops*, males were slightly heavier than females from age class 1 onward, although sexual dimorphism was only significant in adulthood. Table 14.3 compares males and females of each age class for the different taxa and indicates where significant differences can be found.

Could there be life history or evolutionary reasons for these differences in mass and size? Take, for example, the vervets in South Africa, where the animals are the heaviest but the least dimorphic. Could this be due to selective pressures unique to females (e.g., fecundity advantage models: Darwin, 1874; Shine, 1988) or sexual selection for larger females via either male choice or to combat male coercion? In Kenya, higher-ranking females gave birth yearly, while lower-ranking females gave birth every other year. Rank does effect

Table 14.3 *Body size dimorphism in adult Chlorocebus*

Taxon	Sex	n	Body mass (kg)		Dimorphism		Body length (cm)		Dimorphism	
			Mean	SD			Mean	SD		
Ch. aethiops	F	41	2.70	0.51	1.49	p < 0.001	33.28	2.02	1.15	p < 0.001
	M	29	4.02	0.68			38.18	2.91		
Ch. pygerythrus (Kenya)	F	93	2.97	0.48	1.46	p < 0.001	36.77	2.91	1.13	p < 0.001
	M	58	4.34	0.64			41.60	2.98		
Ch. pygerythrus (South Africa)	F	165	4.09	0.66	1.39	p < 0.001	37.18	3.30	1.10	p < 0.001
	M	104	5.69	0.73			40.81	3.56		
Ch. sabaeus	F	64	3.54	0.81	1.60	p < 0.001	34.67	2.57	1.17	p < 0.001
	M	90	5.67	0.74			40.41	3.08		

Notes: "Adult" consists of dental ages 7 and 8. Dimorphism is (mean weight [kg] in adult males) / (mean weight [kg] in adult females).

priority of access to food and may have some relationship to size, but we are unable to determine from our data whether higher-ranking females are also heavier females. Previous work in both wild Kenyan vervets and captive West African savanna monkeys suggesting that fecundity and infant survival are more influenced by female rank than body size (Cheney et al., 1981; Fairbanks & McGuire, 1984) makes this unlikely unless female body size could mediate female rank more significantly in South African than in the other populations. Observations suggest that rank is relatively inflexible in savanna monkeys (e.g., Cheney et al., 1981; Bramblett et al., 1982; Horrocks & Hunte, 1983), however recent work has suggested that male rank, at least, may be influenced by resident females (Young et al., 2017). Intersexual selection via male mate choice is not unknown in cercopithecids, but is often tied to visual cues not observed in savanna monkeys (e.g., large sexual swellings; Higham et al., 2009) or chemosensory cues (Drea, 2015), but not with adult female size. Direct and indirect evidence of male coercion via infanticide does vary across savanna monkey populations, primarily depending on habitat characteristics and demography (Fairbanks & McGuire, 1987; Isbell et al., 2002), but less energetically expensive counterstrategies to male coercion are likely to work in such cases and have been observed in savanna monkeys, including concealed ovulation (Andelman, 1987), male friendships (Keddy Hector & Raleigh, 1992), female–female bonds and cooperation (Cheney, 1983), and reliance on rank to both mitigate coercion efforts by males (Keddy, 1986; for a review, see Smuts & Smuts, 1993) and perhaps increase the rank and reproductive access of preferred males (Raleigh & McGuire, 1989; Young et al., 2017).

An alternative explanation would be that selection is acting on both sexes in South Africa. Animals in South Africa live in the coldest part of the *Chlorocebus* range. Some of the animals routinely experience nighttime low temperatures that are below freezing. An assessment of size in local populations of vervets in South Africa indicated that low nighttime cold temperatures were more important than access to human food in determining body size (Pampush, 2010). The results of the Turner et al. (2018) models also support the role of low temperatures as a selective force affecting body mass and limb length even over the influence of human impacts. Other studies have indicated that vervets in South Africa routinely sun-bathe on cold mornings (Danzy et al., 2012), and there are clear social adaptations to offset thermoregulatory stress (McFarland et al., 2015). Cold, therefore, has had a profound effect on these animals by both altering behavior and perhaps leading to increased mass due to increased fat reserves. Although the growth patterns observed by Turner et al. (2018) suggest that these size differences occur early and that different adult phenotypes are accompanied by divergent growth patterns, the question remains as to whether these adult differences in body mass and limb length are the direct result of acclimatization, developmental plasticity, or selection.

In St. Kitts and Nevis, on the other hand, there is very pronounced sexual dimorphism. Male growth in this population, with comparatively rapid growth in all traits during late develop-ment, leads to a particularly large adult male size relative to females that merits closer investigation (Turner et al., 2018). Interestingly, green monkeys do not have the vivid blue scrota of most other populations of savanna monkey. In the majority of taxa, the inten-sity of the color increases as animals age and might be a signal of maturity and ability to reproduce (Gartlan & Brain, 1968; Isbell, 1995; Gerald, 2001; Gerald et al., 2010). All savanna monkeys in the *Ch. sabaeus* taxon, however, have lighter-colored scrota, while this blanching in adulthood is most pronounced among Caribbean green monkeys. Perhaps the increased degree of sexual dimorphism is com-pensation for the lightened color of the scrota, representing some kind of trade-off in signaling regarding male quality (e.g., Badyaev et al., 2002; Dunn et al., 2015).

It should also be noted that the St. Kitts and Nevis sample of *Ch. sabaeus* analyzed in the Turner et al. (2018) study is from the Caribbean, and not ancestral green monkey populations in West

Africa. These animals were introduced to the Caribbean islands of St. Kitts, Nevis, and Barbados in the seventeenth century. Previous work has shown that this population has significantly larger teeth and crania compared to West African samples, presumably due to a general increase in body size compared to mainland forms (Ashton & Zuckerman, 1950, 1951; Ashton et al., 1979), and that the St. Kitts and Nevis population shows marked genetic homogeneity in comparison to mainland West African populations (Warren et al., 2015; Svardal et al., 2017). Although measurements from the International Vervet Research Consortium sample from The Gambia were too incomplete to be included in the larger analyses of Turner et al. (2018), Wilcoxon signed rank tests comparing both body mass and body length between Gambian adult males and females to those from St. Kitts and Nevis suggest that adult females from The Gambia are actually significantly *larger* than those from St. Kitts and Nevis (body mass: W = 2233, p < 0.001; body length: W = 1579, p < 0.001; Figures 14.1 and 14.2), while males remain similar in size (body mass: W = 749, p = 0.52; body length: W = 221.5, p = 0.07). This appears to contradict the conclusions of Ashton and colleagues that Caribbean *Ch. sabaeus* has been driven toward a larger body size to which the larger cranium and teeth may be scaled; this instead suggests that adult females have perhaps experienced a reduction in body size since crossing the Atlantic. Without increased sampling and complete measures from West African green monkeys, it is unclear whether this comparison of size within *Ch. sabaeus* is a sampling artifact and whether the size patterns noted in the larger analysis of Turner et al. (2018) are characteristic of green monkeys generally or only of the island forms in particular.

In all, this unprecedented sampling of growth across the genus *Chlorocebus* has raised more questions than answers. Our results suggest that the major taxonomic divisions often recognized between populations in this genus do carry some significant differences in both adult phenotype and developmental patterns, but also that local and relatively recently diverged populations – as in Kenyan and South

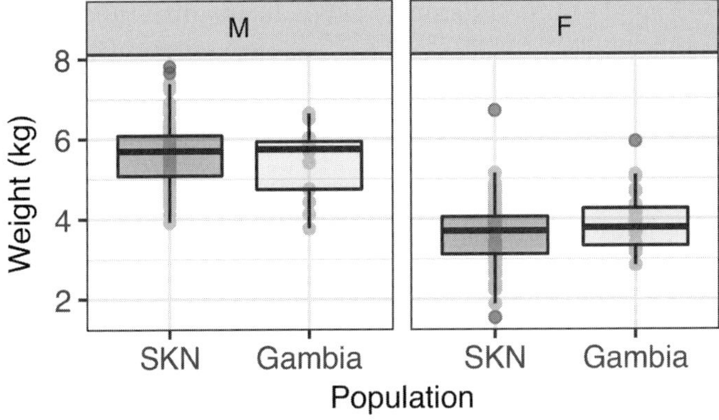

FIGURE 14.1 Mean differences in body mass (kg) between *Ch. sabaeus* populations in The Gambia and St. Kitts and Nevis (SKN) by sex.

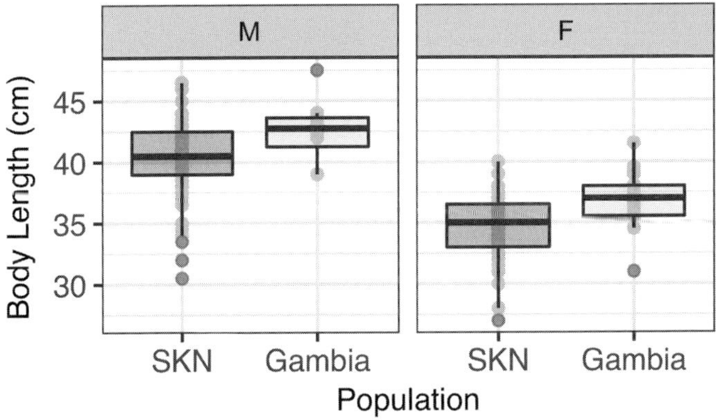

FIGURE 14.2 Mean differences in body length (cm) between *Ch. sabaeus* populations in The Gambia and St. Kitts and Nevis (SKN) by sex.

African *Ch. pygerythrus* – can also vary dramatically in size and patterns of growth. These population differences may reflect unique responses in each to ecogeographic, socio-sexual, or biomechanical selective pressures that may emerge via either natural selection or due to developmental plasticity. Have the two East African taxa converged on a similar weight due to common ecological conditions? How are the divergent morphological and growth patterns seen within the East African and South African *Ch. pygerythrus* populations reached, and do these differences serve an adaptive function? Is the extreme dimorphism of *Ch. sabaeus* due to more intense male–male competition than other savanna monkey populations, and is it linked to a trade-off for the loss of blue scrota as a potential sexual signal? What are the genetic systems that underlie these traits, and can they give us an idea of how these growth and adult phenotypes function and are related? Although our research has answered many questions concerning the life histories of savanna monkeys, we look forward to continuing our research in order to answer these new challenges.

15 The Social and Thermal Competence of Wild Vervet Monkeys

Richard McFarland, Peter Henzi, and Louise Barrett

As long-lived, slowly reproducing animals, primates face numerous ecological challenges to their survival and successful reproduction. The majority of primates live in groups, an adaptation that is widespread in the animal kingdom and can provide improved predator defense, food acquisition, and access to mating opportunities, all of which can contribute to an individual's fitness. Compared to other animal taxa, however, primates are intensely social, spending a significant amount of time forming and maintaining social relationships within their group (Dunbar, 1991). There is evidence to suggest that such relationships are evolutionarily adaptive: more socially integrated individuals experience improved rates of survival and reproductive success (Silk et al., 2009; Schülke et al., 2010; McFarland & Majolo, 2013; McFarland et al., 2017). One argument for the adaptive value of social relationships is that they help offset the inevitable costs of group living, which manifest in terms of increased competition for resources, such as food, safe spatial positions, and mates. The patterning of social life thus represents the negotiation of individual needs within constraints imposed by others. This, in turn, is argued to have selected for a high degree of developmental plasticity and behavioral flexibility among the anthropoid primates in particular.

In addition to these social challenges, climatic variability is also known to exert strong selective pressures on ecology, behavior, and physiology, and has similarly been argued to select for plasticity in the form of developmental norms of reaction, as well as individual behavioral flexibility. Most notably, it has been argued that selection for the ability to cope with and respond to rapid and extensive environmental change can explain patterns of hominin evolution and the marked flexibility of humans compared to other animals (Potts, 1998; Maslin et al., 2015). Understanding the scope and limits of primates' ability to cope with environmental variability is thus the focus of much socioecological research that, broadly speaking, attempts to answer the question of how primates solve their ecological problems with respect to the opportunities and constraints of social life. Such questions are becoming ever more pressing given the impact that global climate change is having on our planet's biodiversity (Thomas et al., 2004; Wiederholt & Post, 2010). Approximately 60 percent of primate species are now threatened with extinction, with 75 percent of primate species experiencing a declining population (Estrada et al., 2017).

Our own research focuses on the behavioral and physiological strategies used by primates to deal with environmental variation, both via changes in resource availability and in terms of direct climatic effects. Specifically, we focus on the thermal physiology of wild vervets (*Chlorocebus pygerythrus*), investigating whether social life compounds or ameliorates the demands made by the thermal environment. Vervet monkeys are ideal model organisms for a study of this type because they are obligatorily social, experience a wide temperature range in arid environments, and manifest a range of specialized behavioral thermoregulatory adaptations. To date, most studies on the thermal physiology of large mammals (which, generally speaking, means non-rodent species) have focused on species that are either solitary or show limited sociality (Fuller et al., 2016). Unlike these species, group-living primates potentially face a compromise between the strategies that promote physiological homeostasis and those that optimize the benefits of group living. Our study is unique, therefore, in allowing us to probe the intersection of our animals' ecological and social strategies, the degree of flexibility they display, and what consequences this holds for survival and reproductive success. This, in turn, provides vital information concerning the long-term viability of vervet populations in the face of ongoing climate change.

Since 2008, we have been studying three groups of vervets on the Samara Private Game Reserve in the Nama Karoo, Eastern Cape, South Africa, combining behavioral, ecological, and body temperature data to investigate individual differences in thermal competence, sociability, and fitness-related traits. Samara is characterized as a high-latitude, semiarid desert, with vervet presence mostly restricted to narrow strips of riparian *Acacia karroo* woodland along non-perennial streams (Pasternak et al., 2013). The region experiences hot, wet summers (November–March) and cold, dry winters (June–August), with minimum and maximum temperatures ranging between –5 and 40°C (McFarland et al., 2014). This region of South Africa is also prone to intermittent periods of drought (Hoffman et al., 2009). Vervets have inhabited the semiarid karoo biome of South Africa since at least the eighteenth century, and despite these extreme environmental conditions, are found at high population densities, with higher than average group sizes compared to other vervet populations in Africa (Pasternak et al., 2013). The rapidity of change in arid-zone thermal environments offers an excellent and feasible opportunity to track the targets of natural selection for thermal competence.

The majority of studies that have explored the effects of environmental variability on primates have tended to focus on the effects of climatic variables on behaviors crucial for survival (e.g., resting, foraging, and social activity: Hill et al., 2003; Campos & Fedigan, 2009; Korstjens et al., 2010; Majolo et al., 2013; McFarland et al., 2014). Importantly, only a few studies have directly collected body temperature measurements from free-ranging primates (Brain & Mitchell, 1999; Dausmann et al., 2004; Mzilikazi et al., 2006; Nowack et al., 2010; Thompson et al., 2014). Although these studies have provided important insights, they have been restricted to few study subjects and short study periods, or lacked detailed simultaneous data on the primates' behavior, feeding ecology, and environment. Moreover, most of these studies used skin or subcutaneous body temperature measurements, which can be significantly affected by ambient temperature, ultimately providing less accurate accounts of the effect of climate on core body temperature. Consequently, the existing literature provides limited insight into the effects of environmental variability on an animal's ability to thermoregulate efficiently. Our study of vervets is the first to use bio-logging technology to collect continuous long-term measurements of core body temperature from multiple individuals in a wild population while simultaneously collecting detailed records of their behavior and ecology (McFarland et al., 2013). Bio-logging has allowed us to determine how efficiently a monkey regulates its body temperature (i.e., daily body temperature means, amplitudes, minima, and maxima) when exposed to environmental and social stressors.

Like all mammals, vervet monkeys are homeothermic, and typically maintain a body temperature ranging between 37 and 39°C (Lubbe et al., 2014). Homeothermy is the ability to maintain a core body temperature within a narrow range when subjected to a wide range of environmental temperatures. This is achieved through a combination of autonomic and behavioral processes. Autonomic processes, involving the activation of pathways in the anterior hypothalamus to regulate the balance of heat production and loss, can be costly in terms of energy expenditure at low temperatures and in terms of water loss (i.e., evaporative sweat) at high temperatures. To help alleviate these costs, animals can also engage in behaviors that help to keep their body warm or cool, such as changing posture or selecting appropriate microclimates. Vervets use various behavioral strategies in their attempts to buffer themselves from environmental variability. During warmer periods, vervets spend significantly more time resting at the expense of feeding (McFarland et al., 2014). To reduce the effect of heat, vervets also seek out shade, hug cool river rocks, retreat into aardvark burrows, and spend time swimming (Figure 15.1). In cooler conditions, vervets can be seen sun-basking or huddling with other group members as they try to maximize heat gain or minimize heat loss, respectively (Figure 15.1). Vervets devote significantly more time to feeding, at the expense of resting, as part

FIGURE 15.1 Behavioral responses of wild vervet monkeys to climatic variability. (a) Drinking, (b) swimming, (c) rock-hugging, (d) resting in shade, (e) grooming, (f) feeding, (g) huddling, and (h) sun-basking.

of their attempt to meet the energetic demands of colder conditions (McFarland et al., 2014).

So far, we have established that vervet monkeys in the Eastern Cape are more prone to cold stress than heat stress (Lubbe et al., 2014; McFarland et al., 2015; Henzi et al., 2017). Specifically, vervets display reduced thermoregulatory efficiency, experiencing lower and increasingly hypothermic body temperatures when temperatures are cold (i.e., the winter months). Moreover, we also see the greatest interindividual variability in thermal competence at this time of year, and these effects become more pronounced as winter progresses and the energetic demands persist. The challenge of the cold is to minimize heat loss to the environment while maximizing heat gain and energy consumption. These problems are exacerbated at night, when vervets retreat to the trees and the risk of predation from land predators is highest (such predators present a significant risk at our study site, where vervets are exposed to black-backed jackal [*Canis mesomelas*], caracal [*Caracal caracal*], and several birds of prey; Pasternak et al., 2013; Ducheminsky et al., 2014). With minimal foraging capability, there is little a monkey can do to buffer itself against cold nighttime temperatures. The only strategy available is to find another group member and engage in huddling. Nocturnal huddling has been observed to be an important adaptation to a range of primate species living in temperate climates (Takahashi, 1997; Ogawa & Takahashi, 2003; Li et al., 2010; McFarland & Majolo, 2013). In line with these previous findings, our study has shown that thermal competence is positively predicted by an individual's number of social partners (McFarland et al., 2015; Henzi et al., 2017), a finding we interpret as reflecting variation in the number of potential huddling partners an animal can call on in cold temperatures. Interestingly, for males, we also found that their tenure length in the troop explains some of the variance in thermoregulatory competence. This suggests that learning and selecting suitable microhabitats at night might also play a role in improving thermoregulatory efficiency (Henzi et al., 2017). In a tangential experimental study of the heat-transfer characteristics of vervet monkey pelts, we found that grooming behavior (Figure 15.1) apparently minimizes heat loss by increasing the pelt's insulative properties (i.e., loft), thus enabling animals to compensate for moderate environmental cooling without increasing demands on their metabolism (McFarland et al., 2016).

During hot conditions, vervets show remarkable efficiency in keeping their body temperatures stable and avoid significant bouts of hyperthermia (Lubbe et al., 2014). It is not surprising that vervets cope better with heat given that, as noted above, they have more strategies for dissipating heat than conserving it. In addition to bio-logging the vervets' temperatures, we also use the same logging technology to measure environmental temperatures and have found that, when vervets seek shade during the hottest part of the day, they select microclimates that can be up to 20 degrees cooler than direct exposure to the sun. Importantly, our vervets typically have regular access to drinking water, which facilitates the sweat production necessary to dissipate heat in these conditions. Although largely a water-dependent species – vervets' geographical distribution is restricted to riverine habitats or artificial water sources maintained by humans – vervets can show remarkable resilience to periods of water unavailability, a circumstance not uncommon within their semiarid, drought-prone habitats. Prior to the beginning of our thermoregulation study (and partly the impetus for it), the vervets experienced a period of intense drought, during which there was very little free-standing water present in the troops' territories (McDougall et al., 2010). As a consequence, there was a marked increase in aggression over access to the little water that was present, despite these being small seepage points that offered little reward. More interesting, however, is what happened at the point when all free-standing water dried up completely. One of our troops was observed to leave their territory on the day after all of the free-standing water dried up, venturing away from the river and along an exposed ridge unoccupied by any other vervet troops. They then returned to the river approximately 750 m from their territory boundary, bypassing the ranges of four other groups and entering an area containing a series of large pools. Our supposition is that the animals were led there by a male immigrant returning to his former territory. The willingness of other group members to follow this male, along with the intense aggression over largely unproductive water sources, suggests that vervets perceive direct access to water as crucial. In another of our groups, the loss of standing water did not prompt immediate movement in search of water. Instead, the group persisted without water for over a month by targeting succulent plants and licking dew from rocks and grass. Eventually, this group similarly followed an immigrant male to the water outside of their home territory. Visits by the two troops were then both regular and frequent, and the monkeys finally began sleeping at the new site. During this period, we observed no drought-related deaths, but it is clear that, had the monkeys not discovered a source of free-standing water, their survival would have been compromised. In this respect, the ability to observe and copy the behavior of animals with a broader experience of the surrounding area reflects another benefit of sociality.

Understanding how social influences on thermoregulation tie into vervet social dynamics is a topic we are now exploring in more detail. The Samara vervets display some interesting differences in their social behavior compared to the classic studies from Amboseli, which may be related to the larger group sizes at our site (Henzi et al., 2013). Specifically, Samara females show clear grooming and proximity preferences for certain individuals, but these patterns are not easily explained by the standard organizing principles of rank and kinship: our females do not groom up the hierarchy, nor do they favor adjacently ranked females (i.e., those who are likely to be kin). There is no relationship between females' spatial proximity to each other and their probability of grooming (i.e., females do not simply groom whichever animal happens to be convenient). We have, however, shown that females use grooming strategically to secure safe spatial positions within the group: animals with larger grooming networks were less exposed to predation risk, from which they benefitted in terms of both reduced vigilance time and increased foraging time (Josephs et al., 2016). It is also apparent, however, that there are large fluctuations in group size over time at our site that reflect variations in climatic conditions – periods of drought result in increased rates of adult and infant mortality – and social patterns may thus vary accordingly. It may be that female social strategies vary in accordance with both group size and prevailing environmental conditions to produce cyclical patterns of variation over time – a possibility our long-term data will allow us to investigate more thoroughly.

To date, the adaptive value of sociability among primates has often been attributed to more sociable individuals being better able to deal with chronic social stress, which has a positive impact on reproduction (Silk, 2007b; Ostner & Schülke, 2018). Collectively, our research on vervets suggests that climatic variability can also present a significant source of stress to primate groups and that more sociable individuals are better equipped to deal with such challenges. If more sociable animals are better able to minimize the metabolic costs of thermoregulation, it is probable that they are not only more likely to survive extreme environmental events, but these savings in maintenance costs mean they also have more energy to invest in reproduction. Our work therefore contributes to the growing body of evidence suggesting that sociability and behavioral flexibility are evolutionarily adaptive traits and are likely to play an important role in promoting the ongoing viability of populations living in highly variable, extreme environmental conditions (Henzi et al., 2009, 2013, 2017; McFarland & Majolo, 2013; Young et al., 2014, 2017; McFarland et al., 2014, 2015, 2017).

16 Novelty-Seeking in Vervets: Developmental, Genetic, and Environmental Influences

Lynn A. Fairbanks

Research with nonhuman primates has provided valuable insights into developmental and genetic influences on the behavioral traits associated with vulnerability for psychopathology for more than 60 years (Suomi, 1982; Kalin, 2004). Early research focused almost exclusively on rhesus monkey models, but in recent years the vervet monkey (*Chlorocebus aethiops*) has emerged as a significant contributor to research in this area. This chapter describes research conducted at the Vervet Research Colony (VRC) on novelty-seeking, a trait of interest for biomedical research that also has relevance for behavioral ecology.

Novelty-seeking is a personality and behavioral trait defined by exploratory activity in response to novel stimuli. In the biomedical sciences, high levels of novelty-seeking have been associated with increased vulnerability for substance abuse, risk-taking and attention deficit hyperactivity disorder (Cloninger et al., 1994; Pawlak et al., 2008). Natural variation in exploration and boldness is also of interest in evolutionary biology and has been related to fitness trade-offs and fluctuating ecological circumstances (Sih & Bell, 2008; Smith & Blumstein, 2008; Blaszcyk, this volume). Research with vervet monkeys has demonstrated that individual differences in exploratory activity and novelty-seeking are multiply determined. The studies described here provide evidence for the effects of early experience, current environment, and genetic variation on individual differences in novelty-seeking.

THE VRC

The VRC was founded in 1975 with an age-graded group of vervet monkeys (now classified as *Chlorocebus aethiops sabaeus*) originally captured from St. Kitts and Nevis, West Indies. Over the years, the colony expanded to its current size of 16 matrilineal social groups. The colony has been managed with the objective of creating a relatively natural social environment within the constraints of captivity. Adult females and their immature offspring remain in their natal group, natal males are moved at four to six years of age, and unrelated breeding adult males are introduced at three- to four-year intervals.

The social groups are housed in large outdoor enclosures with attached indoor shelters. Each enclosure has approximately 150 m^2 of ground area with one or two large elevated platforms and multiple perches and climbing structures.

EFFECTS OF EARLY EXPERIENCE WITH THE MOTHER ON NOVELTY-SEEKING

Our interest in the effects of early maternal care on response to novelty began in the early years. Analysis of behavioral interactions of 50 mother–infant dyads studied at the colony between 1980 and 1987 revealed that maternal behavior varied along two independent dimensions, labeled Protectiveness and Rejection (Fairbanks & McGuire, 1987). The Protectiveness factor loads highest on mother initiates contact and restrains her infant, while the Rejection factor loads highest on mother breaks contact and rejects the infant. Infants of Protective mothers spend less time more than 1 m away from her in the first six months of life.

The first formal novelty-seeking test at the colony was conducted in January 1986 (Fairbanks & McGuire, 1988). In order to accommodate the expanding population, two new enclosures had been constructed that were connected to the original enclosures by temporary tunnels. At the time the tunnels were opened, the groups contained 22 juveniles between the ages of one and three years, all of whom had been observed as infants with their mothers. All of the juveniles in both groups had never been outside of the enclosure where they were born. The new enclosures were comparable in size and shape to the existing enclosures, but they differed markedly in vegetation, with each containing tall grass, bushes, and two small trees.

When the tunnels from the home enclosure into the new enclosures were opened, group members showed signs of interest, excitement, and apprehension. Juveniles were among the first to enter the tunnels. Their latency to enter was not related to their mothers' current responses to the tunnels, but could be predicted by their early experience. Juveniles who had more protective mothers as infants were more cautious and had significantly longer latencies to enter the tunnels compared to juveniles with less protective mothers.

One of the uncontrolled variables in this study was the shared genes between mothers and their juvenile offspring. It is possible that anxious, protective mothers would have hesitant, cautious offspring, independent of early experience. To address this issue, we took advantage of a social circumstance that caused all mothers to increase their protectiveness: the presence of new adult males in the group (Fairbanks & McGuire, 1987). Adult males have been replaced at three- to four-year intervals in the colony, resulting in approximately 30 percent of infants experiencing increased maternal protectiveness and group tension in the first six months of life. In 1987, a new novelty challenge test was conducted, comparing the responses of juveniles who were born in years with new males versus those born in years with resident males (Fairbanks & McGuire, 1993). Comparison of early mother–infant relationships for the 44 subjects verified that the juveniles born in new male years had experienced a higher level of early maternal protectiveness than the juveniles born in resident male years. The novelty stimuli were new upright metal food bins that were introduced to replace the original wooden bins. The monkeys responded to the new food containers with interest and excitement, but also with alarm. Comparison of the two groups of juveniles revealed that juveniles born in new male years had significantly longer latencies to approach the novel food bins compared to juveniles born in resident male years. This study also verified that the juveniles' latencies to approach were not correlated with their mothers' latencies. These results support an effect of variation in maternal protectiveness on juvenile novelty-seeking, above and beyond genetic influences.

In recent years, there has been a growing appreciation of gestational and neonatal effects on offspring developmental trajectories in humans and other mammals (Barker et al., 2002). In the vervet colony, mothers vary in age, experience, weight, and social status, and infants of mothers in relatively poor condition for reproduction (very young, very old, or underweight) had higher rates of neonatal mortality compared to mothers in average or prime reproductive condition (Fairbanks & McGuire, 1984). Marginal mothers were more rejecting to and spent less time in contact with their surviving infants. A recent analysis of the behavior of the surviving offspring of marginal mothers indicated that they showed considerable resilience in social approach and social play as juveniles, but they were significantly below average in novelty-seeking (Fairbanks & Hinde, 2013). This suggests that developmental programming in the womb and during early lactation can have long-term effects on behavioral responses to novel and challenging situations.

DEVELOPMENT OF THE VRC PEDIGREE AND QUANTITATIVE GENETIC ANALYSES

By 2000, the VRC had grown to 16 stable matrilineal social groups with descendants of the founding females in their third to seventh generation. Collaboration with Jeff Rogers and Timothy Newman at the Southwest Foundation led to the development of a paternity panel that allowed the construction of the seven-generation pedigree for the colony (Newman et al., 2002) and the opportunity to explore genetic influences on complex social traits like novelty-seeking.

In 2003, the methods of measuring novelty-seeking were refined and expanded to include an annual assessment of all animals in the 16 matrilineal groups (Fairbanks & Jorgensen, 2011). In the first year, the novelty stimuli were large wading pools added to each outdoor enclosure for enrichment during the hot summer months. Statistical genetics methods were used to measure heritability of responses in the colony pedigree. Analysis of 452 subjects indicated that approximately 50 percent of the variance in latency to approach the new pools (h^2 = 0.52) could be attributed to genetic influences.

At that time, my research group had identified a repeat polymorphism in the candidate gene for the dopamine D4 receptor (DRD4) in the VRC in the same region that has been associated with novelty-seeking and attention deficit disorder in humans (Faraone et al., 2001). When this candidate gene variant was included in the statistical genetics analysis, the results indicated that the DRD4 variant was significantly associated with latency to approach the novel objects, accounting for 13 percent of the population variance, with a residual heritability of 47 percent (Bailey et al., 2007).

NOVELTY-SEEKING INDEX

Between 2003 and 2007, the novelty test was repeated once per year for all colony animals. Attention was paid to selection of novel stimuli that were not too boring or too frightening – stimuli that aroused interest and excitement, but that the large majority of group members would eventually approach within the 30-minute test session. In order to avoid habituation to new objects, predator-like objects were selected as novelty stimuli for 2006–2007. The two primary measures – latency to approach and time spent near the stimulus – were highly correlated in all years and were combined into a single Novelty-Seeking Index. This Index showed a high degree of consistency of individual differences across the five test years (intraclass correlation coefficient = 0.73) and significant heritability (h^2 = 0.42).

It should be noted that all of the tests described here involve free choice, not inescapable novelty. Individuals were tested in their home enclosures and all were free to stay where they were and ignore the stimulus. This is in contrast to inescapable novelty tests that remove an animal from its home environment and confine it in a relatively small and threatening space. Inescapable novelty tests have a long history of being used to measure individual differences in emotional reactivity and anxiety (Kalin, 2004), while free-choice novelty tests tap into individual differences in the tendency to seek, approach, and explore novel situations.

Research with the Novelty-Seeking Index demonstrated that it was a reliable, trait-like, and valid measure of individual differences in response to free-choice novelty, with a significant genetic contribution. There were consistent age differences, with juveniles and adolescents (two to five years of age) having the highest scores and adult scores declining with age. Interestingly, no sex differences were found in any of the tests, nor were there any effects of dominance rank.

PROXIMATE MECHANISMS

The hypothalamic–pituitary–adrenal (HPA) axis is one of the systems activated in response to the perception of danger and a likely candidate to influence novelty-seeking behavior. The high degree of responsiveness of this system made it difficult to study in the colony situation when the act of capturing an animal for blood collection would cause a sudden increase in circulating cortisol. Fortunately, a method for measuring average cortisol release over the preceding one to three months by extracting cortisol accumulation in hair was developed and validated for nonhuman primates (Davenport et al., 2006). Hair samples were collected from colony animals during the annual veterinary exam, assayed in collaboration with Mark Laudenslager, and compared with novelty-seeking scores for adult females in the prior two years. The results indicated that females with chronically high cortisol levels, as measured in hair, were more inhibited and cautious in approaching novel and potentially dangerous objects and had significantly lower novelty-seeking scores. Hair cortisol levels accounted for 7 percent of the variance in scores, indicating that it is one of the factors that influence novelty-seeking behavior.

The role of the HPA axis in novelty-seeking was also supported by an analysis of cerebrospinal fluid from wild vervets in Awash National Park in Ethiopia as part of a larger trapping program

(Phillips-Conroy et al., 1994; Jolly et al., 1996). The free-ranging monkeys were captured when they voluntarily entered baited drop traps over a one-week period. Following capture, each monkey was anesthetized, weighed, and measured, age was estimated, and samples of blood and cerebrospinal fluid were collected. Of the 17 monkeys captured in the first three days of trapping, seven reentered the traps and were captured again one or more times. In other studies, animals that reenter traps are considered to be higher in novelty-seeking and impulsivity than animals who avoid being trapped after their first experience (Higley et al., 1996). Comparison of the levels of three monoamine metabolites in cerebrospinal fluid in the wild vervets revealed that 3-methoxy-4-hydroxyphenylglycol (MHPG) levels were significantly lower in the monkeys who returned to the traps (Fairbanks et al., 1999). MHPG is a metabolite of norepinephrine and an indicator of activity in the HPA axis. Thus, this result is consistent with the findings in captivity that individuals with lower levels of HPA axis activity are more likely to be less cautious and higher in novelty-seeking.

STRESSFUL ENVIRONMENTAL CHANGE

The next stage in the history of the VRC involved a major move of the entire colony to a newly constructed facility at the Wake Forest School of Medicine Primate Center in 2008. The move began with a cross-country road trip followed by a three-month quarantine period, adjustment to a new physical environment, seasonal weather conditions that periodically required confining social groups to the indoor quarters, and management practices that involved more frequent capture and manipulation. As expected, samples collected six months after the move reflected significantly increased levels of cortisol in the hair of colony members (Fairbanks et al., 2011a). Free-choice novelty tests conducted in the first two years following the move demonstrated marked reduction in novelty-seeking behavior for adolescent and adult animals (Fairbanks et al., 2011b). Individual differences in novelty-seeking scores continued to show evidence of significant genetic influences ($h^2 = 0.40$), but genetic correlations conducted within and between the two environments indicated that different genes were involved. Within each environment, the genetic correlation across years was high (rhoG > 0.80), indicating a close correspondence in the genes influencing novelty-seeking within environments. In contrast, the genetic correlation between the two environments was relatively low (rhoG = 0.34) and not significantly different from zero. This result suggests that the effects of environmental stress are not simply additive and provides evidence that different physiological and genetic systems are involved in influencing this behavioral trait under low- and higher-stress conditions.

FUTURE RESEARCH OPPORTUNITIES

The development of new genetics resources for vervets in the VRC and in the wild creates the opportunity to increase our knowledge of the genes and neurophysiological systems that influence variation in novelty-seeking and other behavioral traits (Jasinska et al., 2013; Huang et al., 2015). The free-choice novelty tests described here have been shown to produce a valid and reliable indicator of individual differences in novelty-seeking, a trait with implications for biomedical science and for understanding the ecology of risk and opportunity. The results presented here should be a reminder that complex phenotypes are indeed complex. Novelty-seeking in vervets was related to early maternal condition, maternal protectiveness, and current environment stress, with differential effects by stage of life and different genetic influences emerging in changing environments. The issue of positive correlations between genetic and environmental influences mediated by maternal behavior, kin support, matrilineal dominance rank, and social learning that inflate estimates of heritability must also be considered. Addressing this complexity and controlling developmental and current environmental factors are likely to help drive the discovery of new genes and physiological systems that influence novelty seeking and other complex biobehavioral traits.

ACKNOWLEDGMENTS

The research described here would not have been possible without the help of a large team of dedicated research assistants, animal care personnel, and colleagues, with particular thanks to Karin Blau, Adrianna Huffington, Glenville Morton, and Matthew Jorgensen. Funding was provided by NSF, NIH, and the Department of Psychiatry and Biobehavioral Sciences at UCLA.

17 Measurement of Novelty-Seeking in Wild Vervet Monkeys

Maryjka B. Blaszczyk

The ecology and evolution of animal "personalities" or temperament traits – individual differences in behavior that are stable and consistent across contexts – has become an active and exciting topic of research within behavioral ecology over the last decade (Dall & Griffith, 2014). Comparative psychologists have studied temperament in nonhuman primates for far longer, with vervets being a key model taxon. The trait "novelty-seeking" – the tendency to actively approach and explore novel objects – has been particularly well-studied. As described by Fairbanks (this volume), many of the genetic, physiological, and environmental factors associated with variation in the novelty-seeking trait have been characterized for captive *Chlorocebus sabaeus*. Although individual differences in novelty-seeking are also expected to be related to life history variables and to have fitness consequences in natural populations, the functional consequences of variation in novelty-seeking in wild populations of vervets have yet to be addressed.

In order to assess the ecological relevance of novelty-seeking, it is necessary to study the trait in wild populations that are exposed to the full range of ecological challenges relevant to fitness, including key factors related to food acquisition and predation pressure. Before such studies can be undertaken, however, the feasibility of measuring the novelty-seeking trait in wild vervets needs to be determined. While novelty-seeking tests are performed under standardized conditions in captive vervets, environmental conditions are heterogeneous in the wild, and the complexity of natural environments means that controlling for environmental context is generally not feasible in the field.

Environmental variation might be expected to influence responses to novel objects, and varying environmental conditions may result in a lack of consistency in individuals' exploratory behaviors across novel object tests.

In addition to an immediate effect of environmental context on the expression of behavior, environment can affect behavioral expression over developmental and generational time frames. Several studies have documented the predictive value of gene × environment interactions in the development of temperament trait phenotypes and/or endophenotypes (such as neurotransmitter or hormone levels) in humans (Reif & Lesch, 2003; Ebstein, 2006) and in nonhuman primates (Suomi, 2003, 2006; Barr et al., 2004; Shannon et al., 2005). Research in captive primates and rodents suggests significant and often pervasive effects of environmental factors during development on behavior later in life, including in novelty-seeking and anxiety-related traits (Fairbanks & McGuire, 1993; Clarke et al., 1995; Laviola et al., 2003; Parker et al., 2007; Maestripieri et al., 2009). Finally, captive environments may exert strong selective pressures that can be quite different from those acting on wild populations (McPhee, 2003), and animals with certain personality types that are adaptive in their natural habitat may not do well in captivity, with selection against these phenotypes occurring under captive conditions (Archard & Braithwaite, 2010).

Since environment can potentially impact variation in novelty-seeking in many different ways, an assessment of the reliability of novelty-seeking measurements in wild populations needs to be undertaken as a first step. In animal temperament studies, researchers assess reliability by calculating the *repeatability* of individuals' behaviors across time or contexts. Repeatability refers to the degree to which individual differences in behavior are maintained across time or across contexts or tests, and significant repeatability is the key criterion of temperament. A primary goal of my study in wild South African vervets (*Chlorocebus pygerythrus*) was therefore to determine whether repeatable variation in responses to novel objects can be found in a wild vervet population.

The site for this study was Soetdoring Nature Reserve, Free State Province, South Africa, and the subjects were two troops of vervets occupying contiguous home ranges in the reserve, one group numbering ~55 individuals and the other ~20 individuals. Together with a team of field assistants, I conducted six different novel object tests on each of the groups between April and November 2012. The first three tests were "group-based" tests, where the aim was to simultaneously expose the entire group to a novel stimulus in the center of a large, open clearing as the group crossed this area. These tests were designed to approximate the standard "home-group novelty test" that has previously been used to measure novelty-seeking in captive vervets (Fairbanks & Jorgensen, 2011). Following these tests, we carried out three "individual-based" tests, following methods used for a comparative study of "neophilia" in wild chacma baboons and geladas (Bergman & Kitchen, 2009). In these tests, individual monkeys encountered novel objects that we had placed in their path while they were traveling on the ground between food patches. Study subjects included all adults and subadults in each group. Juveniles were only individually recognizable for the Small group, and juveniles are thus not included in this study.

For the first group-based test, I chose for the novel stimulus two inflatable boats placed side by side, based on their broad similarity to a "wading pool," the object that was used in a study of the behavioral genetics of novelty-seeking in captive vervets (Bailey et al., 2007). This stimulus, as well as materials set up around the object to facilitate accurate recording of test responses (an "inner perimeter line" of ski rope marking out the area within 1 m of the object, an "outer perimeter line" marking out an area within 14 m of the boats, and four tripods holding digital video recorders set up outside this area), proved too threatening a stimulus for wild vervets. Vervets in both groups showed increased levels of vigilance when encountering the test and actively avoided crossing the outer perimeter line. A single juvenile male in the Large group was the only individual to approach to 1 m of the stimulus.

For the second test, the novel object was smaller and more naturalistic – an arrangement of exotic, nontoxic potted plants with brightly colored flowers. The plants elicited a wide range of responses among the monkeys: 58 and 40 percent of non-juveniles approached to within 1 meter of the plants in the Large and Small groups, respectively, and the monkeys' exploratory behaviors included inspection, sniffing, touching, handling, and biting.

For the third group-based test, we used three separate "complex" or multicomponent novel objects, spaced 5 m apart – a bell jar filled with plastic vegetables, an ornate candleholder strung with bath toys, and a toy pirate ship. This test successfully elicited variable behavioral responses from animals in the Large group, with 41 percent of non-juveniles and several juveniles approaching and interacting with the objects. In contrast, only one juvenile male from the Small group approached any of the objects. The lack of responsiveness in this test within the Small group may have been affected by their activity preceding the test, as the group was able to inspect the objects from afar for several hours during their long late-morning rest in a grove of trees on the edge of the testing area before crossing the clearing. Whatever the reason for the discrepancy between the Large and Small groups' responses to this test, no such differences were seen in the other novel object tests, and group identity was not a significant predictor of novel object responses in any analysis of novelty-seeking scores.

For the subsequent individual-based tests, the novel objects used were a large plastic grasshopper, a bright yellow and red toy scorpion, and a realistic-looking gray toy lizard. I chose these objects for their presumed ecological salience to the vervets. The aim of these tests was to capture the first trial for each individual with each of the three stimuli, and tests were conducted in a fixed sequence. For the grasshopper and scorpion tests, behavioral responses were similar to those observed in the group-based plants test, with a portion of individuals in each group sniffing, touching, handling, and biting the novel object. Responses toward the lizard were qualitatively different: most individuals avoided this stimulus, and very few approached to within 1 m. A few monkeys (mostly juveniles) gave snake alarm "chutters" on seeing the lizard. In general, behavioral responses toward the lizard stimulus were similar to those observed when monkeys encountered snakes in their habitat.

Three of the six tests – the plants, grasshopper, and scorpion tests – yielded qualitatively similar behavioral responses and sufficient interindividual variation to be included in the analysis of repeatability of novelty-seeking behavior (Blaszczyk, 2017). I coded behavioral responses to each of the three novel objects from video recordings in the same way. For all first trials with an object, I measured an individual's time spent inspecting the object from within 1 m, the number of times they sniffed the object, the number of "short touches" (manual contact lasting ≤ 1 second), and the duration (in seconds) of longer bouts of touching or handling the stimulus. I combined all sniffing, touching, and handling into a composite "interaction" score, and I assigned a qualitative novelty-seeking score from 0 to 8 based on a combination of inspection time and interaction score. Novelty-seeking scores were positively skewed, with many minimal scores and few high scores. I used a generalized linear mixed model (GLMM) approach to calculate the repeatability of novelty-seeking behavior (Nakagawa & Schielzeth, 2010). Novelty-seeking scores were significantly repeatable across the three tests (link scale r = 0.37, confidence interval = 0.16–0.59, p = 0.003), meeting the criterion of a temperament trait (Blaszczyk, 2017).

I also explored the effects of sex, age class (subadult versus adult), object type, and group on novelty-seeking scores using a GLMM with negative binomial "type 1" (quasi-Poisson) parameterization with the R package glmmADMB (Fournier et al., 2012; Skaug et al., 2013). I included sex, age class, and novel object type, as well as an age × sex interaction as fixed effects. Several females that were classified as subadults at the beginning of the study gave birth toward the end of the novel object testing period and were thereafter classified as adult females. All of the fixed effects in the full model were significant. In the case of the object variable, scores were significantly higher in the grasshopper test than in the plants and scorpion tests. Age class and sex had significant main effects on predicting exploration scores, and there was also a significant interaction between age and sex class. A post hoc multiple comparison test revealed significant differences in novelty-seeking scores between adult females and subadult females and between adult females and subadult males, but not between other age/sex classes (Blaszczyk, 2017).

It is interesting to compare my findings with those from novelty-seeking studies in captive vervets. First, wild vervets in this study were probably generally less responsive toward novel objects than their captive counterparts. The objects used in my study were similar to those used in captive vervet studies. The inflatable boats used in the first test are similar to the wading pool used by Bailey et al. (2007). While the majority of captive vervets approached this "large and potentially threatening novel object" (p. 25), the rubber boats I used proved too threatening a stimulus for wild vervets. Unbeknownst to me when I selected the toy grasshoppers, scorpions, and lizards for my individual-based tests, these are also similar to the "predator-like objects" used in later studies of novelty-seeking in captive vervets (plastic tarantulas, small plastic alligators, and cloth snakes: Fairbanks et al., 2011a; Laudenslager et al., 2011). Laudenslager et al. (2011) reported that 86 percent of female vervets aged 3–18 years approached to within 1 meter of both a cloth snake and a plastic tarantula, and a further 10 percent approached one of these two stimuli. In contrast, the majority of wild vervets avoided the lizard stimulus, and the most common response to the scorpion stimulus was to glance at it for a second or less when passing close by it.

The pattern of very low novelty-seeking scores for wild vervet adult females in particular has not been found in studies of captive vervets. In my study, adult females' novelty-seeking scores were significantly lower than those of subadult males and subadult females. Adult females' scores were also lower than those of adult males, and although this difference was not significant in post hoc tests, the model showed a significant main effect of sex when controlling for age. By contrast, while all captive studies have found an effect of age on novelty-seeking, they have not found significant differences between the sexes, or age by sex interactions (Bailey et al., 2007; Fairbanks et al., 2011). This difference in findings is noteworthy, as sex differences in novelty-seeking have been found in a number of animal taxa (Schuett et al., 2010), including humans. Meta-analyses of human studies find men to have, on average, higher levels of "sensation-seeking" ("the desire to pursue novel or intense experiences"; Cross et al., 2013, p. 1) or "excitement-seeking" than women (Byrnes et al., 1999; Cross et al., 2013) across a diverse range of cultural groups (Costa et al., 2001). The demographic distribution of novelty-seeking among wild vervets therefore parallels the distribution of analogous variation in humans more closely than does variation in novelty-seeking among captive vervets. These results suggest that research using vervet monkeys to model human temperament differences may benefit from including wild vervet populations.

18 Causes of Variation in the Static Allometry of Morphological Structures: A Case Study with Vervet Monkeys

Rafael L. Rodríguez, Tegan J. Gaetano,
J. Paul Grobler, and Nelson B. Freimer

Vertebrates offer an apparent exception to a very general pattern in functional and evolutionary morphology. The general pattern is that male genitalia commonly have near-standard sizes in spite of variation in body size (Eberhard et al., 1998; Eberhard, 2009). In other words, genitalia have negative static allometries. Static allometry describes how structure size scales on adult body size with the slope (b) of log–log regressions (Huxley, 1932). Structures that scale proportionally to body size are isometric ($b = 1$), structures that grow extravagantly large in large individuals but disproportionately small in small individuals have positive allometries ($b > 1$), and structures that vary little in size relative to body size have negative allometries ($b < 1$). Male genitalia exhibit strongly negative allometries in many animal groups, with the notable exception of vertebrates (Eberhard, 2009).

The vertebrate exception to the pattern of negative allometries for male genitalia may be due to the relatively small number of vertebrate species that have been studied with regard to genitalic allometry (Eberhard, 2009). Alternatively, it may be due to a confounding factor involved in the allometry of animals that may continue to grow after sexual maturity, because measuring individuals of different adult ages may confound static allometry with ontogenetic allometry and overestimate b (Eberhard, 2009).

Additionally, the vertebrate exception to negative genitalic allometries may fit with the observation that sexually selected traits, such as ornaments and weapons used in competition for mates, scale very differently relative to the bearer's body size. Some extreme ornaments and weapons, such as male antlers and horns in deer and beetles, scale with strongly positive allometries (Kodric-Brown et al., 2006; Emlen, 2008). But other structures, such as the crest ornaments of some tyrant birds, vary in near-perfect proportionality with body size (Cuervo & Møller, 2001). And yet other ornaments (and genitalia) have near-standard sizes regardless of the body size of the bearer (Eberhard et al., 1998; Cuervo & Møller, 2001; Bonduriansky, 2007a; Eberhard, 2009; Schulte-Hostedde et al., 2011). This variation in the allometry of sexually selected traits requires explanation (Bonduriansky, 2007a; Schulte-Hostedde et al., 2011).

Here, we summarize recent research on vervet monkeys, *Chlorocebus aethiops* (Primates Cercopithecidae) that addresses: (1) the vertebrate exception to the pattern of negative genitalic allometries; and (2) the broad range of variation in the allometry of different traits (Rodríguez et al., 2015b, 2015c).

AGE AND THE STATIC ALLOMETRY OF MALE GENITALIA IN VERVET MONKEYS

In animals like vertebrates, age may be a confounding factor in allometric studies because the body or some body parts may continue to grow in adults. If, for instance, genitalia were to reach full size at an intermediate or late adult age while the body were to stop growing at an early adult age, then genitalia might appear to have steeper b values when assessed among adults of different ages than among adults of the same age. Thus, taking measurements from adults of different ages could overestimate b, forcing isometry or positive allometry depending on the true value of b. If so, there should be a positive relationship between adult growth rates and b estimates across different structures, and estimates of b calculated without accounting for adult age should be steeper than estimates that account for such variation (Eberhard et al., 2018).

We tested these predictions with vervet monkeys (Rodríguez et al., 2015b). We studied two subspecies – *Chlorocebus aethiops sabaeus* in St. Kitts and Nevis and *Chlorocebus aethiops pygerythrus* in South Africa – as part of a broader survey by the International Vervet Research Consortium. Adult males were anesthetized, measured, and returned to the wild within 30 minutes (Grobler & Turner, 2010; Rodríguez et al., 2015b). The males spanned three age categories of sexually mature adults – young adults, adults, and older adults – according to their patterns of dental eruption (Cramer et al., 2013; Rodríguez et al., 2015b). We distinguished between sexual structures (canine and penis length and testes volume) and nonsexual structures (length of the body, head, upper arm, lower arm, thigh, lower leg, sternal notch–pubic symphasis, and girth of the chest, upper arm, and thigh). We estimated b with ordinary least squares (OLS) $\log_{10}$–$\log_{10}$ regression of structure size on the length of the lower leg as a proxy for body size (this measurement had the best-defined landmarks and thus offered the best b estimates; Kilmer & Rodríguez, 2017). For testes volume, we used the $\log_{10}$ of the cubic root of the volume measures. We described growth rates with the slope of OLS regressions of structure size ($\log_{10}$-transformed) on adult age.

We found that, in vervet monkeys, all body parts continued to grow in adults, but some parts grew more than others (Figure 18.1). Sexual structures in particular showed high growth rates (on average, over twice as high as nonsexual traits) (Figure 18.1) (Rodríguez et al., 2015b). Furthermore, there was a positive relationship between the growth rate of a structure and b estimates (Figure 18.2) (Rodríguez et al., 2015b).

It seems noteworthy that, although growth rates were very similar in the two vervet subspecies (Figures 18.1 and 18.2), the b–growth rate relationship seemed to vary, being steeper for *Ch. a. sabaeus* in St. Kitts and Nevis than for *Ch. a. pygerythrus* in South Africa (Figure 18.2). This difference in the b–growth rate relationship

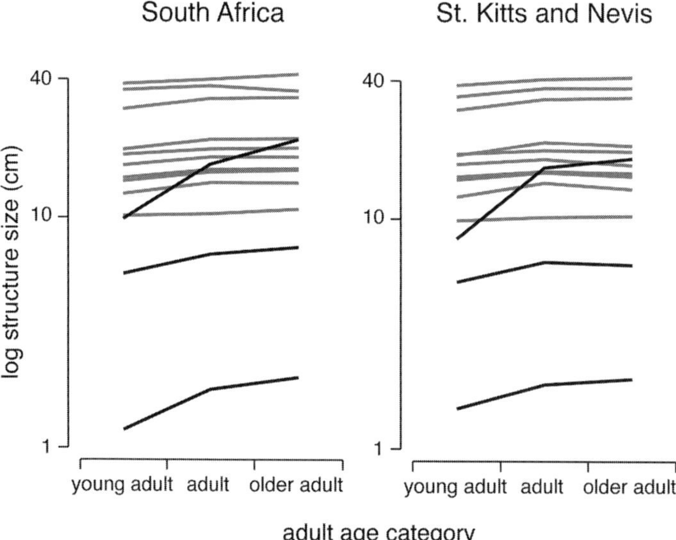

FIGURE 18.1 Growth of different structures over adult ages in two subspecies of vervet monkeys. Black lines: sexual traits (top lines: testes; middle lines: penis; bottom lines: canine length). Gray lines: nonsexual traits (see text). Note y-axes are in log scale.

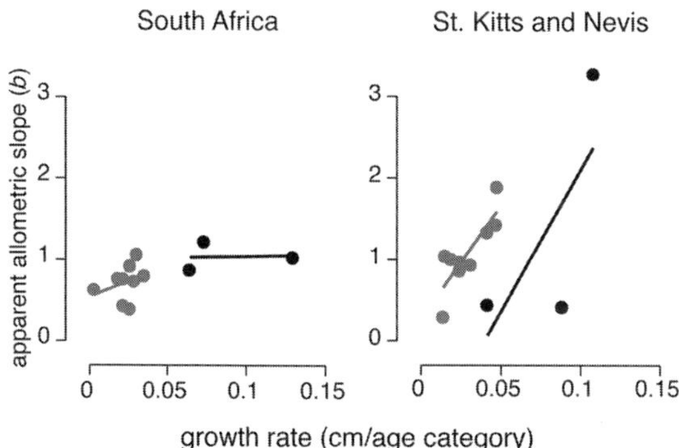

FIGURE 18.2 Positive relationships between apparent allometric slopes (estimated without taking adult age into account) and rate of growth over adult ages in two subspecies of vervet monkey. Gray symbols and lines: nonsexual traits; black symbols and lines: sexual traits.

was only marginally significant (Rodríguez et al., 2015b), but it suggests that there may be geographic variation in the developmental processes that underlie the expression of structure size relative to body size.

Finally, estimates of b calculated without accounting for adult age were higher than estimates accounting for adult age for nearly all traits in both subspecies. For the penis, estimates obtained without accounting for adult age were $b = 0.87$ for *Ch. a. pygerythrus* in South Africa and $b = 0.44$ for *Ch. a. sabaeus* in St. Kitts and Nevis, whereas estimates obtained accounting for age were $b = 0.58$ for *Ch. a. pygerythrus* in South Africa and $b = -0.31$ for *Ch. a. sabaeus* in St. Kitts and Nevis (Rodríguez et al., 2015b).

These results bring vervet monkey genitalia in line with the predominant pattern observed in other animal groups (Eberhard et al., 1998; Eberhard, 2009). It appears that mixing adult ages does seem to bias estimates of the allometric slope for animals that grow after reaching the adult stage. The vertebrate exception to negative genitalic allometries may thus be due to overestimation of b, and it therefore seems profitable to extend our test to other vertebrates.

VARIATION IN THE STATIC ALLOMETRY OF DIFFERENT STRUCTURES IN VERVET MONKEYS

We took advantage of the above data to test a hypothesis proposed to explain the broad range of variation that has been observed for the allometry of different sexually selected structures (Rodríguez et al., 2015c). The proposed explanation involves an interplay between the form of selection (stabilizing versus directional) and body size-related differences in the net benefit of structure size increase (Eberhard et al., 1998, 2009; Bonduriansky & Day, 2003; Bonduriansky, 2007a). The rationale for this hypothesis is that stabilizing selection on structure size will favor negative allometries – because standard structure sizes are favored regardless of variation in body size (Eberhard et al., 1998, 2009) – but that the effect of directional selection may depend on whether its strength varies with body size (Bonduriansky,

2007a; Eberhard et al., 2009). In other words, directional selection that equally favors larger structures for individuals of all body sizes should result in isometry, but if larger individuals benefit to a greater extent from bearing larger structures, then directional selection should result in positive allometry; and if it is the smaller individuals that benefit more greatly, then directional selection should result in negative allometry.

One way to ask about body size-related variation in the net benefits of structure size increase is to assess how condition-dependent different structures are. The rationale for this approach consists of two steps of reasoning: (1) condition-dependent production of structures is expected to evolve for structures that are so costly that only individuals in very good body condition can afford to produce them; and (2) producing costly structures should be relatively cheap for larger individuals (cf., Rowe & Houle, 1996; Bonduriansky, 2007a; Shingleton et al., 2007; Emlen, 2008). This hypothesis therefore makes two predictions: (1) structures under stabilizing selection should be likely to have negative allometries (lower values for b; Rodríguez & Al Watliqui, 2012; Rodríguez et al., 2015a); and (2) more highly condition-dependent structures that are under directional selection should have more positive allometries (higher values for b; cf., Bonduriansky, 2007b; Rodríguez et al., 2015a).

With our vervet monkey data, we focused on testing for a positive relationship between b and the level of condition dependence for different structures. This is a simplistic test because it assumes that all structures are under directional selection. We estimated condition for each male vervet, considering individuals that were relatively heavy for their size to be in better condition than individuals that were light for their size (i.e., they were carrying more muscle and fat; cf., Hunt et al., 2004). We used the residuals of an OLS regression of mass on body length (Schulte-Hostedde et al., 2005). We calculated the level of condition dependence of each structure as the correlation between individual condition and structure size. This measure asked

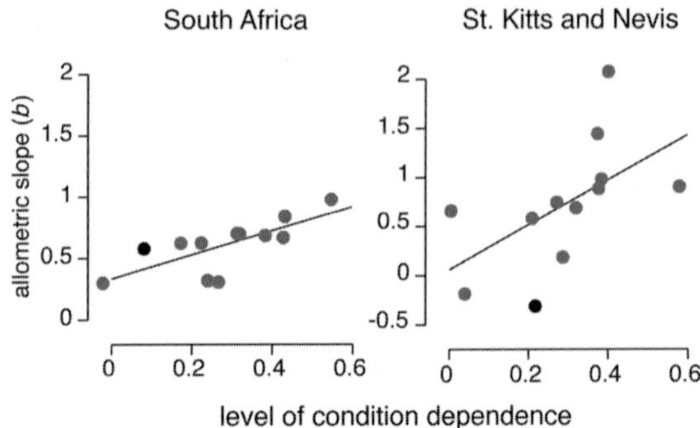

FIGURE 18.3 Relationships between allometric slopes and the level of condition dependence of different structures in two subspecies of vervet monkey.

whether relatively heavy or light males produce larger or smaller structures, independent of body size.

We then asked whether b varied with the level of condition dependence of each trait. We found that b increased with the level of condition dependence in both subspecies (Figure 18.3). There was variation between structure types in this relationship: the penis and the testes had low values for b and condition dependence (black symbols in Figure 18.3 for the penis), but the testes had high values for b and condition dependence (Rodríguez et al., 2015c). Most of the nonsexual traits had intermediate values for b and condition dependence, except for measurements of girth (e.g., upper arm girth), which had $b \approx 1$ but the highest levels of condition dependence (Rutenberg et al., 1987).

The b–condition dependence relationship also seemed to differ between the two subspecies: it was steeper for *Ch. a. sabaeus* in St. Kitts and Nevis, which also showed a broader range of variation in b (Figure 18.3). However, this difference in steepness was not significant (Rodríguez et al., 2015c).

We interpret these results as support for the hypothesis that variation in the static allometry of sexual traits is explained (at least in part) by an interplay between the form of selection and body size-related differences in the net benefit of structure size increase. This support is tentative, however, for a number of reasons. First, we assumed directional selection favoring larger structures (so that lack of support would be inconclusive). Second, our test should be replicated with a broad comparative approach across different animal groups. Also, our test relied on variation among sexual and nonsexual structures to address the problem of variation in sexual allometries – this, however, is a strength, because it argues that the hypothesis proposes a general explanation about the evolution of allometry. In spite of these caveats, our results fits with diverse studies – on morphological and behavioral traits (Rodríguez & Al-Wathiqui, 2012; Rodríguez et al., 2015a), as well as on experimental manipulation of individual condition (Bonduriansky, 2007b) – that provide support for the hypothesis.

In conclusion, vervet monkeys penises fit with a very broad pattern in functional and evolutionary morphology, wherein male genitalia show negative static allometries. This pattern offers insight into the functional design of genitalia – in spite of their showiness, male genitalia in vervet monkeys do not appear to have been selected as indicators of body size or condition or as forceful structures (Eberhard et al., 1998; Eberhard, 2009; Rodríguez et al., 2015c). Additionally, the relationship between allometric slopes and the level of condition dependence that we find in vervet monkey structures offers (in interaction with the form of selection on structure size) a general explanation for the evolution of drastically different allometries in sexual traits (Bonduriansky, 2007a; Eberhard et al., 2009).

ACKNOWLEDGMENTS

The research summarized here was supported by NIH grant R01RR0163009.

Part VI **Ethnoprimatology**

19 Ethnoprimatology and Savanna Monkeys

Trudy R. Turner, Christopher A. Schmitt, and Jennifer Danzy Cramer

The signs are everywhere – "Do not feed the monkeys." "Beware of the monkeys." "Feeding monkeys is signing their death warrant." Vervets are found at every tourist lodge and destination in South Africa. Special garbage cans have been devised that, supposedly, monkeys cannot get into. Guards are found everywhere food may be present. There is a constant battle between people and primates for food and territory. There are people who would be happy to eradicate all vervets; there are people who believe that vervets can be trained to feed at select stations and not throughout a housing estate or neighborhood; there are still others who would happily share with the monkeys. There exists tension not just between monkeys and people, but also between one person and another; tensions between farmers and gardeners and wildlife advocates. And these battles occur everywhere in the animal's range. Human–primate interactions form the core of ethnoprimatology.

ETHNOPRIMATOLOGY

Humans and nonhuman primates have interacted as long as there have been humans and nonhuman primates. Primates, especially what Allison Richard (1985) would call "weed" species – species that inhabit the ecological zone shared with humans – are drawn to the riches associated with human habitation. Food resources are easily accessed with less energy expenditure. And while humans may harass or even kill primates, this threat is probably no greater than those faced anywhere in their territory from other predators. Francis Thackery (e.g., see Gommery et al., 2008) has suggested that papionines are an index species for early hominin finds in southern

Africa. What this indicates is that early species of baboons probably shared territory with early humans from their first appearance. Since baboons also share territory with savanna monkeys, it is not unreasonable to assume that early forms of savanna monkeys also shared habitat with early humans. It is clear why primates would favor human habitats, but why would humans allow primates to be so close? Anyone who has spent time in the field near primates has a ready answer – primates are the extra eyes in the night. Baboons and savanna monkeys can be extremely vocal in the presence of a leopard or a lion. They will provide an early warning system for anyone in the area. Thus, early humans and primates were, in effect, affiliates of each other in a predator-filled landscape – perhaps trading access to food or garbage for warning against predation.

Many early studies of primates took place in parks and reserves or areas that were considered pristine examples of the wild. But even these had a human presence. Parks have lodges; people drive through them. Other people live on the borders of parks. This human presence was mostly ignored during early behavior studies. It was not until Leslie Sponsel (1997) coined the term "ethnoprimatology" – the interface between human and primate ecology – that serious work began on the topic. Since that time, ethnoprimatology has used both ethnographic and primatological methodologies to understand the integrated relationships between humans and primates. Ethnoprimatologists have addressed topics such as the cultural conceptions of human groups that share the primate environment (Cormier, 2002, 2003; Riley, 2007; Fuentes & Hocking, 2010; Riley & Priston, 2010; Fuentes, 2012) and the effects of cohabitation in a series of locations, including Bali (Fuentes et al., 2005; Engel et al., 2006; Loudon, 2006; Lane et al., 2010) and Sulewesi (Lee & Priston, 2005; Riley, 2006, 2010). Researchers have also addressed pathogen transfer (Jones-Engel et al., 2001), parasitology (Loudon, 2006; Fish et al., 2007), and conservation (Fuentes & Wolf, 2002; Lee, 2010). Most of the work in ethnoprimatology has been done on conflict in locations where primates currently reside (Lee & Priston, 2005; Riley,

2007; Hill & Webber, 2010) and efforts at the remediation of this conflict through sanctuaries and rehabilitation facilities (Kessel & Brent, 2001; Schoene & Brend, 2002; Reimers et al., 2007; Wimberger et al., 2010a, 2010b; Guy et al., 2011, 2012, 2013; Trayford & Farmer, 2013). Recent work on ethnoprimatology, however, focuses on the fact that humans and primates are almost always found together, and this can best be understood through a collective mixed methodology using tools from both ethnography and primatology in order to understand these interactions (Dore et al., 2017; McKinney & Dore, 2018).

ETHNOPRIMATOLOGY AND SAVANNA MONKEYS

Rehabilitation Centers

Vervets living in South Africa provide a test case for the utility of sanctuaries and rehabilitation centers. Historically, in South Africa, vervets, along with baboons, leopards, caracals, and jackals, were considered vermin and could be shot on sight. On the other hand, many individuals keep young vervets as pets, even though this practice is illegal (Grobler et al., 2006). Vervets that are problem animals, confiscated pets, or victims of poisoning, shooting, or traffic accidents often come under the care of one of the 20 registered primate rehabilitation centers in the country (Schoene & Brend, 2002; Guy et al., 2013).

In South Africa, it has been estimated that between 900 (Trayford & Farmer, 2013) and 3000 (Grobler et al., 2006) vervets live in either sanctuaries or rehabilitation centers. As opposed to sanctuaries, where animals are provided refuge and care with no intent of future release, rehabilitation programs are facilities that "train inadequate individuals in skills which allow them to survive with greater independence" and "provide former captive individuals the experiences and training to survive and reproduce successfully in the wild when released into their appropriate habitat" (Hannah & McGrew, 1991; Yeager, 1997). Because of the quality of care animals receive and the resistance to euthanasia, rehabilitation centers are often crowded and underfunded (Guy et al., 2013). Some of these centers have begun programs to establish troops of animals with the goal of eventual release back into the wild. Young animals are introduced to adult animals in a structured series of stages, slowly acclimating and assimilating vervets into cohesive groups. Troops coalesce over a considerable period of time. Several releases have been accomplished with varying but often limited success. The limited success of some of these releases led to an evaluation of rehabilitation centers and their release methods by a number of researchers (Wimberger et al., 2010a, 2010b; Guy et al., 2011, 2012, 2013). Although there have been several surveys of primate centers and their intake, troop formation, and release strategies (e.g., see Trayford & Farmer, 2013), there has been virtually no research on animal behavior or stress while in captivity at rehabilitation centers, although recently there has been work on cortisol in research colony management, which may prove useful to rehabilitation centers (Novak, 2003). Vigue (2008) examined vervets at a rehabilitation center in South Africa, obtained fecal samples to assess cortisol levels, and conducted behavioral observations on animals in the facility during the early phases of rehabilitation.

When animals are first brought into a facility, they are quarantined for several weeks. This is followed by an introductory phase where animals are introduced to other vervets. The initial phases of the rehabilitation process in particular can be stressful to individual animals and may cause a physiological response, which can be measured by both cortisol levels and the presence of abnormal behaviors. Previous research has produced inconsistent results when evaluating cortisol response to varying rearing conditions in primates. Early studies comparing peer-reared and isolation-reared primates have reported higher basal cortisol levels in peer-reared monkeys or no difference among the groups (Meyer & Bowman, 1972; Sackett et al., 1973). More recent analyses comparing the fecal cortisol concentrations between captive and semi-captive rehabilitant monkeys and free-ranging monkeys found the lowest cortisol levels in the free-ranging population (Franceschini et al., 1997). However, some studies (Clarke & Mayeaux, 1992; Shannon et al.,

1998) indicate that singly housed, socially isolated savanna monkeys had the lowest measures of fecal cortisol. Shannon et al. (1998) attribute the discrepancies in the activity of the hypothalamic–pituitary–adrenal axis to the type of stressful experience, feeding regimen, and diurnal variation.

While average weekly cortisol levels in the quarantine condition in Vigue's (2008) study were higher than in the introductory condition, these levels were small and not significant. The prediction that cortisol levels would be highest at the beginning of the study and then decrease over time was evidenced only in the quarantine setting. Previous research indicated that there was an overall decline in cortisol in primates that became more familiar with their surroundings or bonded with a new social partner or attachment figure (Mendoza et al., 2000; Rukstalis & French, 2004; Reimers et al., 2007). However, cortisol levels in the introductory setting varied little over the course of the study.

Because social behaviors, both agonistic and affiliative, and nonsocial behaviors, such as withdrawal and stereotypy, are known to influence cortisol output, the correlation between cortisol levels and the amount of time spent engaged in social and nonsocial behaviors was analyzed (DeVries, 2003; Rukstalis & French, 2004). In both the introduction and the quarantine settings, there was no significant correlation between cortisol production and time spent in affiliative, aggressive, defensive, solitary, or abnormal behavior.

The most dramatic finding in Vigue's (2008) study was the reduction of aggressive behaviors in the introductory setting over time. By week 2, aggressive behaviors had reduced substantially. An examination of time budgets for all animals indicated that most social time was spent in affiliative behaviors, and most aggressive behaviors after week 1 were evidenced by a single male and a single female. These findings correlate with previous research on the socialization of baboons and vervet monkeys that revealed increased affiliative behaviors and decreased agonistic interactions over time (Clarke & Mayeaux, 1992; Kessel & Brent, 2001).

Although the introduced vervets did not have a normal model of behavior prior to socialization, they were provided with more space and access to social partners than in their previous housing in the quarantine phase. The change in housing and social environment may have contributed to the decline in abnormal behaviors. Although the levels of abnormal behaviors among vervets in the quarantine condition declined over six weeks, the perseverance of these behaviors may be related to the deleterious consequences of single housing of primates (Kessel & Brent, 2001). The negative consequences of singly housing primates is well documented; however, in order to avoid disease transmission and risk to the health of humans and animals, rehabilitant primates must be quarantined in single cages for some length of time before socialization. Studies that provided enrichment programs to animals where tactile contact between conspecifics is not feasible found that visual and auditory contact based on cage arrangement was useful in reducing abnormal behaviors. Nonsocial programs that offered feeding and cage enrichment with objects such as ropes or toys were effective at reducing abnormal behavior and constituted an important part of an environmental enrichment program for monkeys that could not be housed socially (Kessel & Brent, 1998; Reinhardt & Garza-Schmidt, 2000).

Vigue (2008) expected that the animals in the free-ranging condition would have the lowest cortisol levels. Although individuals in the wild, free-ranging group in this study lived with conspecifics and were free to roam in a more naturalistic environment, their home range was fragmented and in close proximity to humans. Thus, the stress of daily human disturbances and harassment may have elevated the cortisol metabolites in the vervets in the suburban setting. Similarly, in a variety of other primate species, human encroachment, environmental disturbances, forest fragmentation, and reduced food quality and availability have influenced the stress response and increased glucocorticoid production (Chapman et al., 2006; Martinez-Mota et al., 2007). This might explain their elevated levels of cortisol, although these results must be viewed with caution. This study, while based on a small sample size, indicates that vervets can accommodate to new living situations and become active members of an arranged community.

Parasite Load

Given the high potential for pathogen exchange between humans and nonhuman primates, human disturbance of primate habitats, and the low host specificity of many primate pathogens, increasing attention is being paid to how human-driven habitat change may contribute to the emergence or reemergence of bidirectional zoonoses, such as HIV/AIDS, Ebola, simian foamy virus, and polio virus, as well as the transmission of gastrointestinal (GI) parasites (see Chapman et al., 2005; Pedersen et al., 2005). In particular, identifying what aspects of the landscape affect infection in wild primate populations is critical for assessing disease risk and human and primate population health.

Although baseline data have been collected for free-ranging populations of the great apes (e.g., *Pan troglodytes*: McGrew et al., 1989; Ashford et al., 1990, 2000; Bakuza & Nkwengulila, 2009; Petrželková et al., 2010; *Gorilla gorilla*: Lilly et al., 2002; Freeman et al., 2004), baboons (e.g., *Papio cynocephalus*: Hausfater & Meade, 1982; Appleton & Henzi, 1993; Hahn et al., 2003; *Papio anubis*: Kuntz & Myers, 1966; Hahn et al., 2003; Hope et al., 2004; Ryan et al., 2012), colobus monkeys (e.g., *Colobus guereza* and *Colobus angolensis*: Gillespie et al., 2005; *Piliocolobus tephrosceles*: Gillespie & Chapman, 2006), howler monkeys (e.g., *Alouatta palliata*: Stoner, 1996; *Alouatta caraya*: Kowalewski et al., 2011), and forest guenons (*Cercopithecus mitis*: Appleton et al., 1994; *Cercopithecus* spp.: Gillespie et al., 2004), very little is known about patterns of GI parasite infection in other primate taxa. In particular, correlates of infection for species with regular contact with humans, such as members of the monkey genera *Macaca* spp. (macaques), *Papio* spp. (baboons), and *Chlorocebus* (savanna monkeys), remain largely unexplored (Fuentes, 2006).

Savanna monkeys are an established reservoir of infectious zoonotic and anthroponotic agents, including GI pathogens (Jones & Scott, 1970; Hahn, 1994; Muriuki et al., 1998; Toft & Eberhard, 1998; Mutani et al., 2003; Legesse & Erko, 2004). The route of exchange of GI parasites between hosts is oral–fecal contact, most often through contaminated soil, water, or food items (Chapman et al., 2005; Nunn et al., 2011). These attributes make savanna monkeys an ideal primate model for research on the ecology of parasite infection across anthropogenically modified environments.

Gaetano and colleagues (2014) surveyed a number of vervets representing 15 troops across their range in South Africa during collections for the International Vervet Research Consortium. Fecal samples were collected in addition to Global Positioning System (GPS) locations of all samples. These locations were plotted against boundary data, human impact data, and other mapping data. Of 78 sampled animals, 55.13 percent (n = 43) were found to be positive for parasites. The highest prevalence of parasite infection was 48.72 percent for strongylids. Prevalence rates for coccidia and *Trichuris* spp. were 3.85 and 7.69 percent, respectively. These percentages approximate those found for wild South African vervets by Wren (2006). Strongylid density ranged from 0 to 267 eggs per gram of fecal sample. The greatest richness of parasites (i.e., all three parasite species present in a single host) as well as the highest density of strongylid infection (267 eggs per gram of sample) occurred along the eastern coastal regions of the country, with *Trichuris* spp. only found at these sites. Strongylids and coccidia were homogeneously distributed among sample sites. Proximity to dense human centers was not a reliable predictor of parasite infection prevalence, density, or richness in wild populations of South African vervet monkeys. Although statistical significance was not reached, strongyle infection densities tended to be highest for sites with intermediate human contact, coded in the Geographic Information System (GIS) as "rural."

Chapman and colleagues (2016) studied the effects of parasites on the social networks of savanna monkeys at their field site in

Uganda. After deworming the population, they found that individuals engaged in less resting behavior, traveled more and had more frequent interactions with other members of the group. As such, they suggest that parasite loads can affect not only growth, but also behaviors that may have evolutionary significance.

An examination of parasites in primates as well as an examination of microbiomes are two of the ways in which primatological studies are moving in new directions. Kerry M. Dore presents a nuanced examination of ethnoprimatology on the island of St. Kitts, examining what has happened since the termination of large-scale sugar plantations on the island. Sponheimer and colleagues discuss the ways in which stable carbon and nitrogen isotopes can be used as measures of contact between humans and nonhuman primates. They compare levels found in vervets and chimpanzees in differing environments and explore how this technique allows researchers to determine interactions between species.

20 Exploring Caribbean Green Monkeys (*Chlorocebus sabaeus*) through an Ethnoprimatological Lens

Kerry M. Dore

This book highlights many reasons why vervet monkeys are an attractive study species. Among the most notable are the species' opportunistic nature and extreme behavioral and ecological flexibility. Due in part to this set of characteristics, a number of these monkeys survived trans-Atlantic journeys on ships transporting slaves and goods from West Africa to the Caribbean and successfully colonized the islands of Barbados, St. Kitts, and Nevis by the mid-1600s (Denham, 1982a, 1987; Dore, 2017a, 2017b). In this essay, I review the data on Caribbean green monkeys in the primatological literature and explain how my work on the relationship between monkeys, colonial histories, and citizen farmers in this region contributes to the growing research approach of ethnoprimatology.

CARIBBEAN GREEN MONKEYS IN PRIMATOLOGICAL LITERATURE

The earliest scholarly publications focused on Caribbean green monkeys do not appear until the mid-1900s. However, the animals were officially declared vermin in Barbados (and a bounty was offered for each monkey killed) in 1682, so it is likely that the animals were present in large numbers on St. Kitts, Nevis, and Barbados by this time (Sade & Hildrech, 1965; see Denham, 1982a, 1987 for alternative possibilities). Father Labat, a French priest who visited St. Kitts in 1700, provides the earliest known report of the animals in the Federation of St. Kitts and Nevis. Labat saw "wild troops roaming the island and actively 'crop raiding,' so much so that crop guards were necessary" (McGuire, 1974, p. 6). The historical frequency of complaints about crop damage by monkeys and the initiation of a bounty support the idea that the animals were thriving soon after they set foot on the islands.

Scientific inquiries into Caribbean green monkeys begin after nearly 300 years of isolation from their African counterparts. This literature starts with Cameron's (1930) comparison of the internal parasite *Subulura distans* in St. Kitts and West African vervet monkeys, and Coyler, Ashton, and Zuckerman's comparisons of the dental abnormalities and skulls of 92 Kittitian vervets compared to West African controls (Colyer, 1948; Ashton & Zuckerman, 1950, 1951a, 1951b, 1951c; Ashton, 1960). The differences observed by these individuals played a role in the designation of Caribbean green monkeys as the subspecies *Cercopithecus sabaeus* (McGuire, 1974). While many scholars and islanders still call these Caribbean primates "vervet monkeys," as *sabaeus*, their official common name is now "green monkey" (Groves, 2005; Wallis, 2013).

Naturalistic studies on the St. Kitts green monkey begin in 1965, with Sade and Hildrech's (1965) general assessment of the monkey population and Poirier's (1972) investigation of the monkeys' social behavior and population dynamics. To date, the most significant research into free-ranging Caribbean green monkey behavior began in 1970 and resulted in the book *The St. Kitts Vervet* (McGuire, 1974). The book contains detailed observations of nine monkey troops in St. Kitts: three groups living in the dense rainforest interior of the island and six troops from the dry, savanna-like southeast peninsula. At this time, these were the two main areas of the island occupied by monkeys (Figure 20.1). Due to the numerous challenges associated with observing monkeys living in the forested interior (McGuire, 1974, pp. 2–3), behavioral studies of Kittitian monkeys in the 1970s and 1980s took place on the southeast peninsula and focused on topics such as interindividual distance, territoriality, general social behaviors, and habitat use (Fairbanks & Bird, 1978; Chapman & Fedigan, 1984; Chapman, 1985, 1987; Chapman et al., 1988). A few

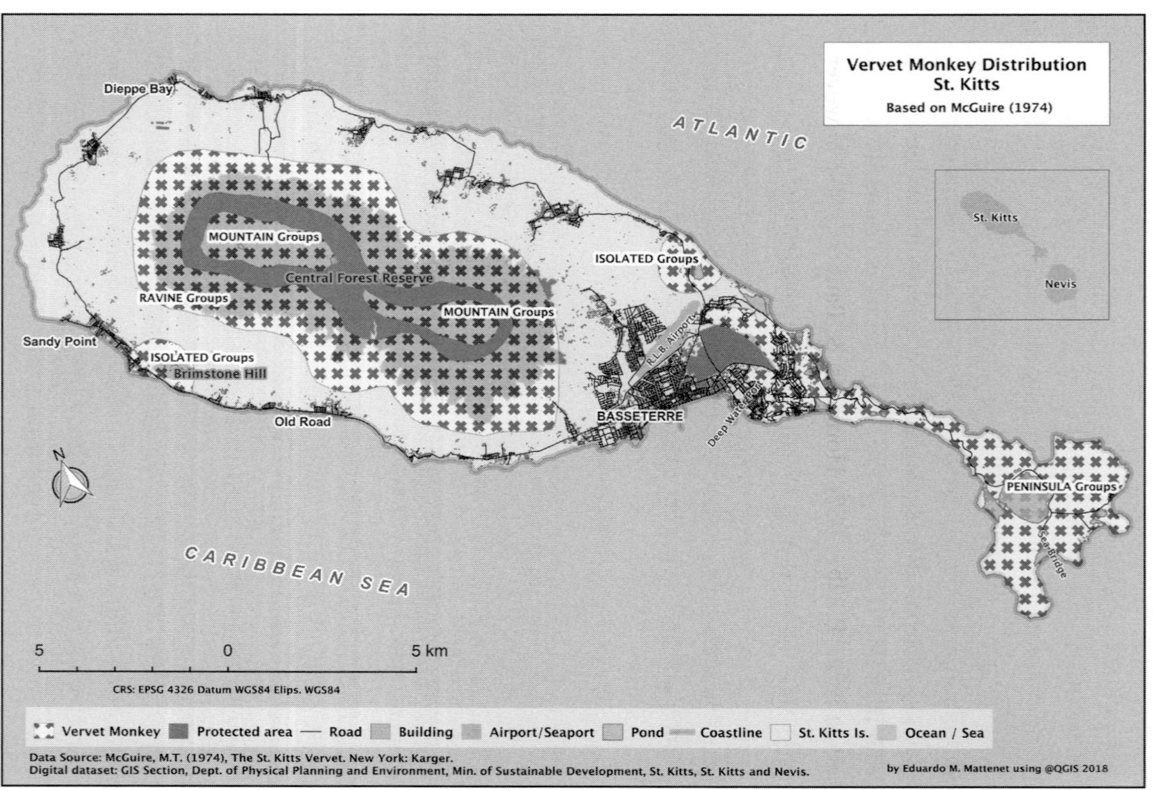

FIGURE 20.1 Location of green monkeys prior to sugar industry closure (reproduced from McGuire, 1974, with permission from S. Karger AG, Basel, Switzerland).

additional studies of St. Kitts green monkeys addressed issues related to demography and population size (Coppinger & McGuire, 1980; Fedigan et al., 1984).

Scholarly publications on Barbadian green monkeys began in the 1980s. Early studies concentrated on the disease status (Blakeslee et al., 1985; Pedersen et al., 1986; Baulu et al., 1987a, 1987b), behavior (Horrocks, 1982, 1984; Horrocks & Hunte, 1983a, 1983b, 1986, 1993; Hunte & Horrocks, 1987), and life history characteristics (Horrocks, 1986) of free-ranging monkeys. Like in St. Kitts, questions about monkey population size began to appear in the literature at this time (Denham, 1982b, 1987; Horrocks & Baulu, 1988); however, in Barbados, more attention was paid to the interaction between monkeys and people. Specifically, these studies focused on the issue of crop damage and food competition between farmers and monkeys (Horrocks & Baulu, 1994; Boulton et al., 1996).

FARMER–GREEN MONKEY INTERCONNECTIONS IN ST. KITTS

In the summer of 2010, I initiated my doctoral research on St. Kitts monkey crop-foraging behavior. My goals were to systematically document the monkeys' crop consumption patterns, farmers' reactions to this behavior, and the historical, economic, and political context of monkey–farmer interactions. My results show that while monkeys' current crop-foraging patterns are predictable in nature, one must consider Kittitian farmers' broader relationships with both monkeys and their environment in order to fully understand the human–nonhuman primate (hereafter primate) interface in St. Kitts.

Ethnoprimatology

Ethnoprimatology, a nascent yet distinct line of research within biological anthropology, provided the framework for my analysis. Exactly which elements of the human–primate interface to investigate in an ethnoprimatological study must be carefully catered to one's particular study species and site. For example, one must consider the needs and wishes of the local government and people, the habituation level of the animals, the relationship between local people and primates, and whether this relationship includes potentially sensitive or controversial topics such as primate hunting, consumption, or pet ownership. When planning my initial research in St. Kitts, I focused on recording patterns of monkey crop damage, farmers' reactions to this damage, and the broader context of this conflict for two main reasons. First, there are no habituated free-ranging monkeys on the island. The animals cannot be habituated; it would only make them more susceptible to human predation, as there is a severe negative local perception of the monkeys. Documenting patterns of crop damage left behind by the monkeys is a means of understanding their behavior indirectly. Second, the local government and agricultural community expressed strong interest in a scientific study on this topic. The presence of in-country collaborators is often a key factor in the successful completion of primate field studies (Jones-Engel et al., 2011).

Methods

My initial fieldwork in St. Kitts was conducted from June 2010 through to August 2011. With help from the St. Kitts Department of Agriculture, I compiled a list of the country's registered farmers and sought permission from a randomly selected subset (one-third, n = 65) of this group to participate in monthly visits in order to document crop loss and in an open-ended interview about their experiences with monkeys. Quantitative data on crop loss were collected using handheld Garmin® Global Positioning Systems (GPS) devices and localized to an individual grid cell within an island-wide half-acre grid system that I created in ArcMap v. 9.3, a Geographic Information System (GIS). Data on nine potential predictors of crop loss were collected for each grid cell/farm. Interviews occurred as the comfort level with the farmer increased; farmers who were more open and communicative were interviewed at the beginning of the study and those who were more reserved were interviewed later after a relationship had been established (see Dore et al., in press, for details on quantitative methods, and Dore, 2017c for a personal account of the study design and analysis).

Results and Discussion

The results of this investigation fall into three interrelated categories: the predictable nature of green monkey crop-foraging behavior; the relationship between Kittitian farmers and monkeys; and the relationship between Kittitian farmers and their environment, specifically the land they cultivate and the broader economic context of Kittitian agriculture. In terms of monkey crop consumption patterns, I developed two models: one that can predict the likelihood of a farm experiencing crop damage by monkeys and one that can predict the severity of crop damage a farm will receive. Using these data, scripts have been created in Excel that allow one to input data on the nine predictive variables for any piece of land in St. Kitts and obtain a value for both the likelihood of experiencing damage and the severity of this damage (once damage has been predicted).

While these quantitative results are informative about monkey crop-foraging behavior at the present time, historical and ethnographic data provide important context about the role of history and politics in this environmental issue. Data from archival research, interviews, and field notes reveal that the closure of the sugar industry, which dominated the island's landscape from the 1600s until 2005, has significantly impacted green monkey movements and their contact with farmers. Prior to 2005, monkeys were restricted to the forest interior, ravines, and southeast peninsula, as the rest of the land was protected by sugar industry infrastructure (crops, equipment, workers, and rangers actively shooting animals to keep them away from the sugarcane). The St. Kitts government, composed of local elected officials since the Federation of St. Kitts and Nevis gained independence in 1983, acquired the sugar lands from European estate owners and ran the industry starting in the mid-1970s. After the industry closed, they allocated small plots of this land to former sugar workers. These farms and farmers were the focus of my study. While these individuals are essentially farming the same land, they no longer have the protection of the estate system keeping the monkeys contained in the forest and ravines. They are thus viewing monkey behavior and witnessing monkey crop damage on a large scale for the first time. Seeing monkeys' human-like qualities so clearly in the last ten years has drastically shifted Kittitian farmers' cultural conceptualization of "monkey" and made it challenging to construct a singular identity for them (Dore, 2013; Dore et al., 2018).

The vast majority of farmers in St. Kitts do not own the land they cultivate; most of the agricultural land is former cane land that is owned by the government and rented or leased to farmers. This land tenure system has created insecurity on the part of some farmers about the future of St. Kitts agriculture, as recent land use changes have favored development and tourism over agriculture and resulted in the relocation of a number of farms that had been cultivated by specific families for generations. Farmers explain that their lack of land ownership also impacts their ability to control crop damage from animals. Many farmers argue that they would be more willing to invest in protecting their farm were they more confident in their ability to cultivate that land for the long term (Dore et al., 2018; Dore, 2018).

These results support some of the core tenets of ethnoprimatological theory. In this investigation, including the human dimension and the perspective that farmers and green monkeys in St. Kitts mutually shape each other's ecologies has exposed the complex interplay between monkey behaviors, farmer behaviors, and farmer perceptions. Themes from animal studies in sociocultural anthropology, a field that influenced the development of ethnoprimatology (Fuentes, 2012), are present in this work, such as the notion that phenomena that cannot be ordered into classes (i.e., those that cross spatial boundaries) are anomalous. In St. Kitts, the monkeys are now "matter out of place" (Douglas, 1966): as they pass through the buffer zone between forest/ravine and farm that was maintained by the sugar industry for hundreds of years, they traverse the nature–culture boundary (being "natural" animals in "cultural" [i.e., agricultural] spaces) and the human–animal boundary (with increased recognition of the monkeys' human-like behavior). Thus, in St. Kitts, green monkey pestilence is as much about the animals' boundary crossing as it is about the economic consequences of their crop raiding (Knight, 2000; Dore et al., 2018). These results show why vervet monkeys are an excellent study species for ethnoprimatological research: their flexible nature enables them to adapt to changes in their habitat and cross spatial and conceptual boundaries, key attributes of primates that shape human ecologies.

Ethnoprimatological research is increasingly interested in how facets of the human–primate interface interact with broader historical, political, and economic contexts to coproduce landscapes (Fuentes, 2010; Jost Robinson & Remis, 2014; Dore, 2018). This case study from St. Kitts adds to the mounting body of evidence in the ethnoprimatological literature on the influence of anthropogenic habitat changes/disturbances on nonhuman primate behavior and ecology (e.g., Knight, 1999; Fuentes et al., 2005; Hardin & Remis, 2006; Sprague & Iwasaki, 2006; Riley, 2007; Campbell-Smith et al., 2010; Hockings et al., 2010). Like in Japan, changes in the historical ecology of St. Kitts have increased the interface between humans and monkeys and drastically increased rates of crop damage (Sprague, 2002). In addition to documenting the means by which land use changes influence human and primate behavior, the ethnographic data collected for this investigation highlight how the political and economic relationship between humans and their environments – in this case, the lack of land ownership on the part of Kittitian farmers and the economic priority of tourism – also influences the human–primate interface. Human–primate conflict in St. Kitts is a serious problem, not only because monkeys are present to a much higher degree in agricultural areas, but also because of the country's insecure land tenure system and concerns about the fate of agriculture itself (Dore et al., 2018; Dore, 2018).

CURRENT AND FUTURE WORK

The results of the ethnoprimatological research described in this chapter have important implications for the management of green monkeys in St. Kitts and the broader Caribbean. At the most basic level, my work documents a drastic increase in the interface between humans and monkeys in St. Kitts in the last 13 years. This increase has important negative consequences, including heightened rates of crop damage, food insecurity, and potential for disease transmission between humans and monkeys. My current work and future plans revolve around developing a monkey management strategy for St. Kitts and initiating a comparative research program to analyze human–primate interconnections on all of the Caribbean islands that are home to green monkeys: St. Kitts, Nevis, Barbados, and, most recently (in the last 20 years, brought via boats from St. Kitts and Nevis), St. Martin/St. Maarten (Brown, 2008).

As monkeys in St. Kitts have no natural predators, islanders have long been arguing that the only appropriate management strategy is a large-scale monkey cull. Prior to any interventions with regard to the monkey population, however, a systematic and scientific estimate of population size is necessary. Over the last five years, I have collaborated with the Food and Agriculture Organization of the United Nations, the Christophe Harbour Foundation, and Ross University School of Veterinary Medicine (RUSVM) on an island-wide green monkey population estimate. My current work with Dr. Christa Gallagher from RUSVM involves estimating the range sizes and troop sizes of two monkey groups from each of four habitat types in St. Kitts: the southeast peninsula, villages, agricultural areas, and the central forest. To determine range sizes, we use Global System for Mobile Communications (GSM) and iridium tracking devices to monitor the movements of one representative adult male from each troop for approximately three to four months. To determine troop sizes, we hire Kittitian monkey trappers (with more than 30 years of experience) to observe each monitored troop for at least 20 hours from within a hide, collecting data on group size and demographics. To date, we have data from two troops from the southeast peninsula, two troops in farming areas, two troops in the forest habitat, and one troop from a village habitat. We are currently tracking and observing the second village troop (see Dore et al., 2014 for our estimate of the southeast peninsula green monkey population). Additionally, in January 2018, I began a camera trapping project in collaboration with the Smithsonian National Museum of Natural History. The 28 cameras currently deployed along transects across the island will provide (among other things) an additional means of estimating monkey population density. Using the telemetry and camera trap data, my existing data on patterns of monkey crop damage, and historical and archival data on the effectiveness of the sugar industry infrastructure at reducing the human–green monkey interface, I will model and test the most effective means to protect agricultural areas from monkeys in St. Kitts.

Over the long term, I will evaluate how the varying geographies, ecologies, and economics of other Caribbean islands impact local human–primate entanglements. Nevis, St. Kitts' sister island, shares much of St. Kitts' early history, including the transformation of the land for sugar production and the subsequent importation of slaves and monkeys. However, the island stopped producing sugar back in the 1940s, and as a result has significantly higher levels of individual land ownership and a long-standing tourism economy. Do Nevisians have different relationships with monkeys as a result? Barbados, which also shares the early history of sugar production, slavery, and green monkeys, is flat, with a significantly larger human population. How do these different geographic and demographic patterns, as well as the fact that the island still produces sugar and has a long history of culling monkeys, impact the human–monkey interface? Finally, St. Martin/St. Maarten has only been home to monkeys for a few decades, making the island a model for what the early colonization of St. Kitts, Nevis, and Barbados may have been like. How do the interactions between monkeys and individuals on this two-nation island differ as a result of the monkeys' novel presence and the fact that the island is entirely dependent on tourism, growing virtually none of its own food? While green monkeys' presence on Caribbean islands has led to very serious environmental concerns, it has also created a host of exciting new opportunities for innovative anthropological research.

21 Vervet Monkeys (*Chlorocebus pygerthrus*), Chimpanzees (*Pan troglodytes*), and Humans (*Homo sapiens*): Studying Interactions Using Stable Isotope Analysis

Matt Sponheimer, James E. Loudon,
J. Paul Grobler, Kimberly Moyer,
and Joseph G. Lorenz

INTRODUCTION

Approximately half of the earth's nonhuman primates are considered threatened by the International Union for Conservation of Nature (IUCN). This is due primarily to anthropogenic influences, including habitat destruction or modification, hunting, and capture for trade (Mittermeier et al., 2012). Interactions between humans and nonhuman primates are often viewed negatively given that they often result in our nonhuman kin being displaced from their natural habitats and/or faced with local extirpation. However, primatologists who focus on human–nonhuman primate interconnections have revealed a diversity of human and nonhuman primate relationships demonstrating that simple positive or negative characterizations are too simplistic (Fuentes, 2012, 2014). Traditionally, these approaches have fused a variety of behavioral observational techniques with cultural anthropological methodologies, including interviews and questionnaires (but see Riley & Ellwanger, 2013).

A few studies have used stable carbon ($\delta^{13}C$) and nitrogen ($\delta^{15}N$) isotope analyses to examine the impacts of humans on sympatric nonhuman primate behavior (Loudon et al., 2007, 2014; Schurr et al., 2012; Schillaci et al., 2014). Stable isotope ratios are expressed as δ values relative to international standards in parts per thousand (permil), as in the following example for carbon isotopes: $\delta^{13}C$ (‰) = $(R_{sample} / R_{standard} - 1) \times 1000$, where $R = {}^{13}C/{}^{12}C$. The $\delta^{13}C$ and $\delta^{15}N$ values of animals reflect the $\delta^{13}C$ and $\delta^{15}N$ values of the foods they consume with some additional fractionation that varies depending on the tissue (e.g., dental enamel, bone, hair) of interest. Most primates consume primarily plant foods, and there are three types of plant photosynthetic pathways (C_3, C_4, and Crassulacean acid metabolism [CAM] plants) that typically, or in many cases always, have different carbon isotopic compositions. C_3 plants include dicot trees, shrubs, and temperate grasses, C_4 plants include most tropical grasses and some sedges, and CAM plants include succulents like cacti and some euphorbias (Kluge & Ting, 1978). The carbon isotope compositions of C_3 and C_4 plants do not overlap (C_4 plants have higher $\delta^{13}C$ values), and therefore the stable isotope compositions of an animal's tissues reveal the relative amounts of C_3 or C_4 plants it consumes (Lee-Thorp & van der Merwe, 1987; Cerling et al., 2006). Nevertheless, among C_3 plants, there exists a considerable degree of carbon isotopic variation. Plants growing underneath a dense canopy have lower $\delta^{13}C$ values than those growing in more open areas due to incorporation of ${}^{13}C$-depleted CO_2 produced by decaying leaves (Medina & Minchin, 1980) and lower light intensities (Ehleringer et al., 1986). As a result, when humans degrade forest habitats, resulting in a loss of canopy closure, the $\delta^{13}C$ values of the remaining C_3 plants tend to increase.

Most naturally occurring plants consumed by primates use C_3 photosynthesis, but throughout much of Africa, people grow C_4 crops (e.g., corn, millet, sorghum, and sugarcane) that are opportunistically consumed by crop-raiding primates. Given the isotopic difference between most wild primate foods (C_3) and the human crops (often C_4) they raid, stable carbon isotope analysis should allow us to trace consumption of human foods by nonhuman primates. Moreover, since the degradation of forest habitats also leads to higher $\delta^{13}C$ values, anthropogenic influences in general would tend to result in higher primate $\delta^{13}C$ values. Nitrogen isotopic compositions may also reveal anthropogenic influences, most notably because crops fertilized with natural fertilizers will typically have higher $\delta^{15}N$ values than other local foods (Bateman & Kelly, 2007) and because consumption of animal food products, which are ubiquitous in human food refuse, also tends to result in higher $\delta^{15}N$ values (Schoeninger & DeNiro, 1984). Thus, all else being equal, one might expect higher $\delta^{15}N$ values to be indicative of anthropogenic influence. However, plant and animal $\delta^{15}N$ values are much more difficult to interpret because they are highly influenced by local and regional nitrogen cycles and are therefore most readily interpreted when good data on local plant $\delta^{15}N$ values are available (see Loudon et al., 2016).

Here, we seek to build upon previous work using stable isotopes to investigate the impact of humans on nonhuman primate $\delta^{13}C$ and $\delta^{15}N$ values (Loudon et al., 2007, 2014; Schurr et al., 2012; Schillaci et al., 2014). We use hair from two primate species, vervet monkeys and chimpanzees, both of which can be found in African "savanna" environments (Wrangham & Waterman, 1981; Pruetz & Bertolani, 2009), and for which there are $\delta^{13}C$ and $\delta^{15}N$ data from environments with both high and low anthropogenic disturbance. Following the above, we expect to find that groups of vervets and chimpanzees with significant anthropogenic influence will have higher $\delta^{13}C$ values. There is also reason to believe that anthropogenic influence might lead to a similar increase in $\delta^{15}N$ values, although we expect that unknown baseline differences in plant $\delta^{15}N$ values (e.g., Craine et al., 2009) will make the central tendencies of high- versus low-disturbance sites difficult to distinguish.

METHODS

We collected 100 hair samples from ten groups of vervet monkeys (*Chlorocebus pygerythrus*) throughout South Africa (Loudon et al., 2014) and four hair samples from a population of vervet monkeys (*Chlorocebus sabaeus*) in St. Kitts. Samples were collected from populations that were known to be significantly impacted by humans and from others where human influence was minimal. Monkeys were baited with corn or fruit and captured in traps designed to minimize the risk of injury to the monkeys and humans and to facilitate administration of tranquilizers (Grobler & Turner, 2010). All field protocols were approved by the Interfaculty Animal Ethics Committee of the University of the Free State and the Institutional Animal Care and Use Committee at the University of Wisconsin-Milwaukee. Hair was collected from the upper shoulder of each individual as a part of a larger study examining vervet monkey population genetics, genomics, phylogeography, and biology (Jasinska et al., 2013). We compared the $\delta^{13}C$ and $\delta^{15}N$ hair values of these vervet monkey groups to those of free-ranging chimpanzee (*Pan troglodytes*) communities from three "savanna" sites (Ishasha and Ugalla: Schoeninger et al., 1999; Fongoli: Sponheimer et al., 2006), three forest communities in Uganda (Kanyanchu, Kanyawara, and Chambura Gorge: Loudon et al., 2016), and three highly anthropogenically influenced sites in Uganda (Miranga Village, Lake Kerere, and Uganda Wildlife Education Center [UWEC]: Loudon et al., 2016). We chose the forest sites we did (rather than sites from Ivory Coast or Cameroon, for instance) because we felt they were most appropriate for comparison with our three anthropogenically influenced sites, all of which are in Uganda.

Hair samples were cleaned with ethyl alcohol, cut into segments with a razor blade, weighed (~700 μg), and placed in tin capsules. Samples were combusted in an elemental analyzer and analyzed for stable carbon and nitrogen isotope abundances using a flow-through inlet system on a continuous-flow isotope ratio mass spectrometer. We determined the degree of anthropogenic disturbance (high versus low) for each vervet group and chimpanzee community based on our observations or published accounts. We used Welch's analysis of variance (ANOVA) for all comparisons between species, groups with different levels of disturbance, and, for chimpanzees, habitat. We examined the effects of species and disturbance level on isotopic compositions using the Fit Model feature (least squares method) in JMP Pro12 with $\delta^{13}C$ and $\delta^{15}N$ as response variables and disturbance (high versus low) and species (vervet versus chimpanzee) as effects

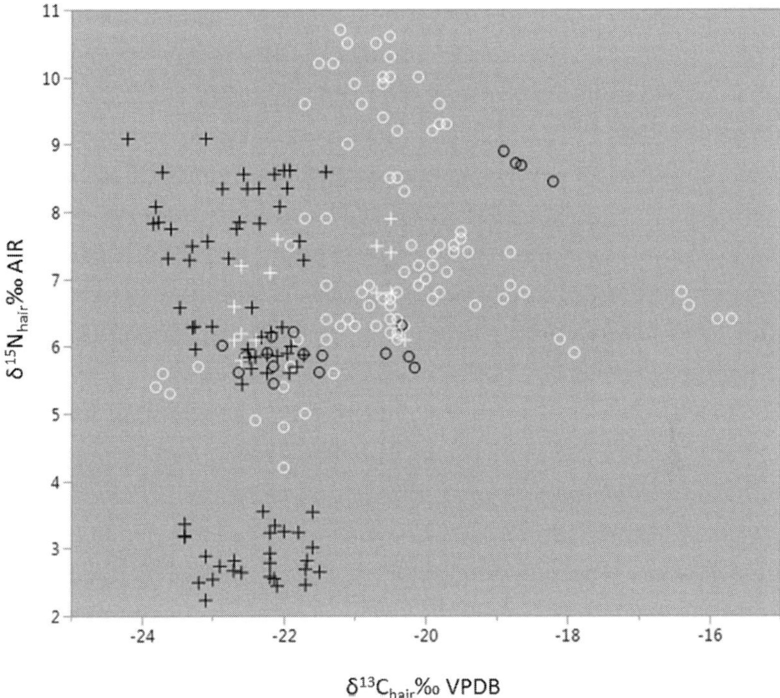

FIGURE 21.1 $\delta^{13}C$ and $\delta^{15}N$ biplot for vervet monkey and chimpanzee populations. Light symbols represent vervet monkeys and dark symbols represent chimpanzees. Circles illustrate high levels of anthropogenic disturbance and crosses represent low levels of or little disturbance. VPDB = Vienna Pee Dee Belemnite.

(interaction terms were included initially). In some cases, data were log transformed or ranked prior to analyses in order to avoid violating assumptions of the tests employed, but as these transformations did not alter the significance of the results appreciably relative to non-transformed values, we present results for the non-transformed data set below. Data used for this paper are available at https://doi. org/10.6084/m9.figshare.7545158.

RESULTS

Figure 21.1 shows the $\delta^{13}C$ and $\delta^{15}N$ values for each vervet and chimpanzee specimen. Welch's ANOVA revealed differences in the mean $\delta^{13}C$ ($F_{1,190.35} = 82.97$; $p < 0.0001$) and $\delta^{15}N$ ($F_{1,170.67} = 35.29$;

$p < 0.0001$) values of vervets and chimpanzees. Among the vervet monkeys, we found significant differences in the $\delta^{13}C$ ($F_{1,24.61} = 14.59$; $p < 0.001$) and $\delta^{15}N$ ($F_{1,47.97} = 6.11$; $p < 0.05$) values of populations living in habitats with high versus low anthropogenic disturbance. Among chimpanzees, we found differences in both $\delta^{13}C$ ($F_{1,20.91} = 16.02$; $p < 0.001$) and $\delta^{15}N$ values ($F_{1,58.94} = 5.77$; $p < 0.05$) for communities living in habitats with high versus low levels of anthropogenic disturbance. We also found that "savanna" and forest chimpanzees (excluding high disturbance sites) have different $\delta^{13}C$ ($F_{1,22.54} = 32.18$; $p < 0.0001$) and $\delta^{15}N$ ($F_{1,63.01} = 54.36$; $p < 0.0001$) values. Two models (one for $\delta^{13}C$ and another for $\delta^{15}N$) exploring the effects of species and disturbance on $\delta^{13}C$ and $\delta^{15}N$ values were significant ($p < 0.0001$), as were the effects species ($p < 0.01$) and disturbance ($p < 0.05$) in both models (these models do not include interaction terms as initial runs revealed no disturbance and species interaction). However, R^2 values were low for both models ($\delta^{13}C$, $R^2 = 0.40$ and $\delta^{15}N$, $R^2 = 0.18$), such that between 60 and 82 percent of the variance in $\delta^{13}C$ and $\delta^{15}N$ remained unexplained. The unexplained variance is probably linked to unknown habitat or behavioral differences between populations.

DISCUSSION

It is not surprising that there are differences in the $\delta^{13}C$ values of vervets and chimpanzees (vervet $\delta^{13}C$ values are higher), especially as some of the chimpanzee sites are from closed environments (e.g., Kanyawara, Kanyanchu) where we would expect a canopy effect. The same mechanism can explicate differences in the $\delta^{13}C$ values of the "savanna" and forest chimpanzee groups. Only a handful of vervet values fall within the range of forest chimpanzees (without high anthropogenic disturbance). In contrast, there are no differences in the $\delta^{13}C$ values of vervets and chimpanzees with high levels of anthropogenic influence ($F_{1,26.81} = 3.22$; $p = 0.084$), suggesting that disturbance in these environments and anthropogenic contexts tends to impart a similar isotopic signature to both primate species. We also found differences in the $\delta^{15}N$ compositions of vervets and chimpanzees and in the $\delta^{15}N$ values of forest and "savanna" chimpanzees, but these are much more difficult to interpret given baseline differences in the $\delta^{15}N$ values of vegetation from place to place (Craine et al., 2009; see Loudon et al., 2016). Consequently, it is difficult to ascribe any ecological or biological significance to these differences in central tendency.

Of particular note, and as predicted, both vervets and chimpanzees show trends toward higher $\delta^{13}C$ and $\delta^{15}N$ values in disturbed habitats. This makes sense as in these contexts disturbance often includes clearing of landscapes, leading to less canopy-driven depletion in the $\delta^{13}C$ values of plants, or the use of human crops, which are typically ^{13}C-enriched C_4 plants like corn, sorghum, or sugarcane. Since both vervets and chimpanzees are expected to consume few C_4 resources under normal circumstances, their consumption would be expected to significantly alter the isotopic compositions of consumers, and this study appears to bear that out. The higher $\delta^{15}N$ values in disturbed habitats make sense, as opening of landscapes, consumption of foods grown with natural fertilizers, and access to greater amounts of animal foods, all of which can happen as a result of human disturbance, may also lead to elevated $\delta^{15}N$ values. Once again, however, this trend in $\delta^{15}N$ values is difficult to interpret, as consumption of anthropogenic resources might also decrease $\delta^{15}N$ values if synthetic fertilizers are used (Bateman & Kelly, 2007) or if legume crops such as peanuts are raided (Schoeninger et al., 1999). Moreover, differences in baseline plant $\delta^{15}N$ values from place to place due to local nitrogen cycling add another complicating factor. The latter problem can be overcome if local plants are sampled in addition to hair, but this was not done during the present study or the other studies from which we derived the isotopic data used here.

Visual inspection of the $\delta^{13}C$ data for both species suggests that values of –21.0‰ or greater are usually indicative of anthropogenic disturbance. Yet, individuals with lower values could also be from disturbed environments. As for the $\delta^{15}N$ data, values that approach

9.0‰ are likely to indicate anthropogenic inputs, at least across the range of environments represented by our samples. This is not an overly useful measure as the vast majority of the specimens sampled fall below this threshold, making $\delta^{13}C$ a much more sensitive indicator of anthropogenic influence, at least in the absence of plant baseline data.

CONCLUSIONS

This study investigated the impacts of species and anthropogenic disturbance on the $\delta^{13}C$ and $\delta^{15}N$ values of two primate species. There is a strong pattern where $\delta^{13}C$ values above –21.0‰ likely indicate an anthropogenic signal. While $\delta^{13}C$ values are not perfect indicators of disturbance, they are quick and inexpensive, and are an especially useful supplement if hair is collected for DNA analyses or other purposes. We believe that stable isotope analysis can be a powerful tool for revealing cryptic feeding behaviors (i.e., crop raiding) and nonhuman primate and human interplays when long-term behavioral studies are not feasible.

ACKNOWLEDGMENTS

We thank the following institutions: the Department of Economic Development and Environmental Affairs, Eastern Cape; the Department of Tourism, Environmental and Economic Affairs, Free State Province; Ezemvelo KwaZulu-Natal (KZN) Wildlife; the Department of Economic Development, Environment and Tourism, Limpopo; the Department of Agriculture, Conservation and Environment, Mpumalanga; the Department of Environment and Nature Conservation, Northern Cape; and the Limpopo Department of Economic Affairs, Environment and Tourism. We are indebted to the veterinarians, landowners, and field assistants involved in this project. This work was supported by grants from the National Science Foundation, the National Institutes of Health, the Fulbright Foundation, the University of Limpopo, and the University of the Free State.

References

Aiello, L.C. & Wheeler, P. (1995). The expensive-tissue hypothesis: the brain and digestive system in human and primate evolution. *Current Anthropology*, 32(2), 199–221.

Akinyi, M.Y., Tung, J., Jeneby, M., Patel, N.B., Altmann, J. & Alberts, S.C. (2013). Role of grooming in reducing tick load in wild baboons (*Papio cynocephalus*). *Animal Behaviour*, 85, 559–568.

Alberts, S.C. (1999). Paternal kin discrimination in wild baboons. *Proceedings of the Royal Society B*, 266, 1501–1506.

Albrecht, G.H. (1980). Latitudinal, taxonomic, sexual, and insular determinants of size variation in pigtail macaques, *Macaca nemestrina*. *International Journal of Primatology*, 1, 141–152.

Albrecht, G.H., Jenkins, P.D. & Godfrey, L.R. (1990). Ecogeographic size variation among the living and subfossil prosimians of Madagascar. *American Journal of Primatology*, 22, 1–50.

Alexander, R.D. (1974). The evolution of social behavior. *Annual Review Ecology, Evolution, and Systematics*, 5, 325–383.

Allal, N., Sear, R., Prentice, A.M. & Mace, R. (2004). An evolutionary model of stature, age at first birth and reproductive success in Gambian women. *Proceedings of the Royal Society of London B: Biological Sciences*, 271(1538), 465–470.

Altmann, J. & Alberts, S. (1987). Body mass and growth rates in a wild primate population. *Oecologia*, 72(1), 15–20.

Altmann, J. & Alberts, S.C. (2003). Variability in reproductive success viewed from a life-history perspective in baboons. *American Journal of Human Biology*, 15(3), 401–409.

(2005). Growth rates in a wild primate population, ecological influences and maternal effects. *Behavioral Ecology and Sociobiology*, 57, 490–501.

Altmann, J., Altmann, S. & Hausfater, G. (1981). Physical maturation and age estimates of yellow baboons, *Papio cynocephalus*, in Amboseli National Park, Kenya. *American Journal of Primatology*, 1, 389–399.

Altmann, J., Schoeller, D., Altmann, S.A., Muruthi, P. & Sapolsky, R.M. (1993). Body size and fatness of free-living baboons reflect food availability and activity levels. *American Journal of Primatology*, 30 (2), 149–161.

Altmann, J., Alberts, S.C., Haines, S.A., Dubach, J., Muruthi, P., Coote, T., Geffen, E., Cheesman, H., Mututua, R.S., Saiyalel, S.N., Wayne, R.K., Lacy, R.C. & Bruford, M.W. (1996). Behavior predicts genetics structure in a wild primate group. *Proceedings of the National Academy of Sciences of the USA*, 93(12), 5797–5801.

Altmann, S.A. & Altmann, J. (1970). Baboon ecology. *Bibliotheca Primatologica*, 12, 1–220.

Amato, K.R., Yeoman, C.J., Cerda, G., Schmitt, C.A., Cramer, J.D., Berg-Miller. M.E., Gomez, A., Turner, T.R., Wilson, B.A., Stumpf, R.M. Nelson, K.E., White, B.A., Knight, R. & Leigh, S.R. (2015). Variable responses of human and nonhuman primate gut microbiomes to a Western diet. *Microbiome*, 3, 53.

Amedee, A.M., Rychert, J., Lacour, N., Fresh, L. & Ratterree, M. (2004). Viral and immunological factors associated with breast milk transmission of SIV in rhesus macaques. *Retrovirology*, 1, 17.

Andelman, S.J. (1985). Ecology and reproductive strategies of vervet monkeys (Cercopithecus aethiops) in Amboseli National Park, Kenya (PhD dissertation). University of Washington at Seattle.

(1987). Evolution of concealed ovulation in vervet monkeys (*Cercopithecus aethiops*). *The American Naturalist*, 129(6), 785–799.

Andersson, M. (1994). *Sexual Selection*. Princeton, NJ: Princeton University Press.

Andres, M., Solignac, M. & Perret, M. (2003). Mating system in mouse lemurs: theories and facts, using analysis of paternity. *Folia Primatologica*, 74, 355–366.

Apetrei, C., Robertson, D.L. & Marx, P.A. (2004). The history of SIVs and AIDS: epidemiology, phylogeny and biology of isolates from naturally SIV infected nonhuman primates (NHP) in Africa. *Frontiers in Bioscience*, 9, 225–254.

Apetrei, C., Gormus, B., Pandrea, I., Metzger, M., Haaft, P., Martin, L.N., Bohm, R., Alvarez, X., Koopman, G., Murphey-Corb, M., Veazey, R.S., Lackner, A.A., Baskin, G., Heeney, J. & Marx, P.A. (2004). Direct inoculation of simian immunodeficiency virus from sooty mangabeys in black mangabeys (*Lophocebus aterrimus*): first evidence of AIDS in a heterologous African species and different pathologic outcomes of experimental infection. *Journal of Virology*, 78, 11506–11518.

Apetrei, C., Gautam, R., Sumpter, B., Carter, A.C., Gaufin, T., Staprans, S.I., Else, J., Barnes, M., Cao, R., Jr., Garg, S., Milush, J.M., Sodora, D.L., Pandrea, I. & Silvestri, G. (2007). Virus-subtype specific features of natural SIVsmm infection in sooty mangabeys. *Journal of Virology*, 81, 7913–7923.

Appleton, C.C. & Henzi, S.P. (1993). Environmental correlates of gastrointestinal parasitism in montane and lowland baboons in Natal, South Africa. *International Journal of Primatology*, 14(4), 623–635.

Appleton, C.C., Krecek, R.C., Verster, A., Bruorton, M.R. & Lawes, M.J. (1994). Gastro-intestinal parasites of the Samango monkey, *Cercopithecus mitis*, in Natal, South Africa. *Journal of Medical Primatology*, 23(1), 52–55.

Archard, G.A. & Braithwaite, V.A. (2010). The importance of wild populations in studies of animal temperament. *Journal of Zoology*, 281, 149–160.

Ashford, R.W., Reid, G.D.F. & Butynski, T.M. (1990). The intestinal faunas of man and mountain gorillas in a shared habitat. *Annals of Tropical Medicine and Parasitology*, 84(4), 337–340.

Ashford, R.W., Reid, G.D.F. & Wrangham, R.W. (2000). Intestinal parasites of the chimpanzee *Pan troglodytes* in Kibale Forest, Uganda. *Annals of Tropical Medicine and Parasitology*, 94(2), 173–179.

Ashton, E.H. (1960). The influence of geographic isolation on the skull of the green monkey (*Cercopithecus aethiops sabaeus*). Part V. *Proceedings of the Royal Society of London. Series B*, 151, 538.

Ashton, E.H. & Zuckerman, S. (1950). The influence of geographic isolation on the skull of the green monkey (*Cercopithecus aethiops sabaeus*). Part I. *Proceedings of the Royal Society of London. Series B*, 137, 212–238.

(1951a). The influence of geographic isolation on the skull of the green monkey (*Cercopithecus aethiops sabaeus*). Part II. *Proceedings of the Royal Society of London. Series B*, 138, 204–213.

(1951b). The influence of geographic isolation on the skull of the green monkey (*Cercopithecus aethiops sabaeus*). Part III. *Proceedings of the Royal Society of London. Series B*, 138, 213–218.

(1951c). The influence of geographic isolation on the skull of the green monkey (*Cercopithecus aethiops sabaeus*). Part IV. *Proceedings of the Royal Society of London. Series B*, 138, 354–374.

Ashton, E.H., Flinn, R.M., Griffiths, R.K. & Moore, W.J. (1979). The results of geographic isolation on the teeth of the green monkey (*Cercopithecus aethiops sabaeus*) in St. Kitts – a multivariate retrospect. *Journal of Zoology*, 188, 533–555.

Atkins, H.M., Wilson, C.J., Silverstein, M., Jorgensen, M., Floyd, E., Kaplan, J.R. & Appt, S.E. (2014). Characterization of ovarian aging and reproductive senescence in vervet monkeys (*Chlorocebus aethiops sabaeus*). *Comparative medicine*, 64, 55–62.

Atkinson, E.G., Rogers, J., Mahaney, M.C., Cox, L.A. & Cheverud, J.M. (2015). Cortical folding of the primate brain, an interdisciplinary examination of the genetic architecture, modularity, and evolvability of a significant neurological trait in pedigreed baboons (genus *Papio*). *Genetics*, 200, 651–665.

Avise, J.C. & Ball Jr., R.M. (1990). Principles of genealogical concordance in species concepts and biological taxonomy. *Oxford Surveys in Evolutionary Biology*, 7, 45–67.

Backhed, F., Ding, H., Wang, T., Hooper, L.V., Koh, G.Y., Nagy, A., Semenkovich, C.F. & Gordon, J.I. (2004). The gut microbiota as an environmental factor that regulates fat storage. *Proceedings of the National Academy of Science*, 101, 15718–15723.

Baden, A.L., Brenneman, R.A. & Louis Jr., E.E. (2008). Morphometrics of wild black-and-white ruffed lemurs (*Varecia variegate*; Kerr, 1792). *American Journal of Primatology*, 70(10), 913–926.

Badyaev, A.V., Hill, G.E. & Weckworth, B.V. (2002). Species divergence in sexually selected traits: increase in song elaboration is related to decrease in plumage ornamentation in finches. *Evolution*, 56, 412–419.

Bailes, E., Gao, F., Bibollet-Ruche, F., Courgnaud, V., Peeters, M., Marx, P.A., Hahn, B.H. & Sharp, P.M. (2003). Hybrid origin of SIV in chimpanzees. *Science*, 300, 1713.

Bailey, A. L., Lauck, M., Sibley, S.D. et al. (2015). Zoonotic potential of simian arteriviruses. *Journal of Virology*, 90, 630–635.

Bailey, J.N., Breidenthal, S.E., Jorgensen, M.J., McCracken, J.T. & Fairbanks, L.A. (2007). The association of DRD4 and novelty seeking is found in a nonhuman primate model. *Psychiatry & Genetics*, 17, 23–27.

Bakuza, J.S. & Nkwengulilia, G. (2009). Variation over time in parasite prevalence among free-ranging chimpanzees at Gombe National Park, Tanzania. *International Journal of Primatology*, 30(1), 43–53.

Baldellou, M. & Henzi, S.P. (1992). Vigilance, predator detection and the presence of supernumerary males in vervet monkey troops. *Animal Behaviour*, 43(3), 451–461.

Barker, D.J.P., Eriksson, J.G., Forsen, T. & Osmond, C. (2002). Fetal origins of adult disease, strength of effects and biological basis. *International Journal of Epidemiology*, 31, 1235–1239.

Barnicot, N.A. & Hewett-Emmett, D. (1971). Red cell and serum proteins of talapoin, patas, and vervet monkeys. *Folia Primatologica*, 15, 65–76.

   (1972). Red cell and serum proteins of *Cercocebus*, *Presbytis*, *Colobus* and certain other species. *Folia Primatologica*, 17, 442–457.

Barnicot, N.A, Jolly, C.J. & Wade, P.T (1967) Protein variations and primatology. *American Journal of Physical Anthropology*, 27, 343–356.

Barr, C.S., Newman, T.K., Shannon, C., Parker, C., Dvoskin, R.L., Becker, M.L., Schwandt, M., Champoux, M., Lesch, K.P., Goldman, D., Sumoi, S.J. & Higley, J.D. (2004). Rearing condition and rh5-HTTLPR interact to influence limbic–hypothalamic–pituitary–adrenal axis response to stress in infant macaques. *Biological Psychiatry*, 55, 733–738.

Barrett, A.S., Brown, L.R., Barrett, L. & Henzi, P. (2010). A floristic description and utilisation of two home ranges by vervet monkeys in Loskop Dam Nature Reserve, South Africa. *Koedoe*, 52, 1–12.

Barton, R.A. & Harvey, P.H. (2000). Mosaic evolution of brain structure in mammals. *Nature*, 405(6790), 1055–1058.

Basckin, D.R. & Krige, P.D. (1973). Some preliminary observations on the behaviour of an urban troop of vervet monkeys (*Cercopithecus aethiops*) during the birth season. *Journal of Behavioral Science*, 1, 287–296.

Bateman, A.S. & Kelly, S.D. (2007). Fertilizer nitrogen isotope signatures. *Isotopes in Environmental and Health Studies*, 43, 237–247.

Baulu, J., Everard, C.O.R. & Everard, J.D. (1987a). Leptospires in vervet monkeys (*Cercopithecus aethiops sabaeus*) on Barbados. *Journal of Wildlife Diseases*, 23, 60–66.

(1987b). The African green monkey (*Cercopithecus aethiops sabaeus*) as a carrier of disease in Barbados. *Laboratory Primate Newsletter*, 26, 2–4.

Bayes, M.K., Smith, K.L., Alberts, S.C., Altmann, J. & Bruford, M.W. (2000). Testing the reliability of microsatellite typing from faecal DNA in the savannah baboon. *Conservation Genetics*, 1(2), 173–176.

Bearder, S.K. (1987). Lorises, bushbabies, and tarsiers: diverse societies in solitary foragers. In B.B. Smuts, D.L. Cheney, R.M. Seyfarth, R.W. Wrangham & T.T. Struhsaker, eds., *Primate Societies*. Chicago: University of Chicago Press, pp. 11–24.

Bell, G. & Koufopanou, V. (1986). The cost of reproduction. In R. Dawkins, ed., *Oxford Surveys of Evolutionary Biology*. Oxford, UK: Oxford University Press, pp. 83–131.

Benefit, B.R. (1999). *Victoriapithecus*, the key to Old World monkey and catarrhine origins. *Evolutionary Anthropology*, 7(5), 155–174.

Benefit, B.R. & McCrossin, M.L. (2001). Craniodental comparisons of *Mabokopithecus* and *Oreopithecus* support and African origin of Oreopithecidae. *American Journal of Physical Anthropology*, 32, 37–38.

Benefit, B.R., Gitau, S.N., McCrossin, M.L. & Palmer A.K. (1998). A mandible of *Mabokopithecus clarki* sheds new light on oreopithecid evolution. *American Journal of Physical Anthropology*, S26, 109.

Berard, J.D., Nuernberg, P., Epplen, J.T. & Schmidtke, J. (1993). Male rank, reproductive behavior, and reproductive success in free-ranging rhesus macaques. *Primates*, 34(4), 481–489.

Bercovitch, F.B., Widdig, A., Trefilov, A., Kessler, M.J., Schmidtke, J., Nuernberg, P. & Krawczak, M. (2003). A longitudinal study of age-specific reproductive output and body condition among male rhesus macaques, Macaca mulatta. *Naturwissenschaften*, 90 (7), 309–312.

Berghänel, A., Schülke, O. & Ostner, J. (2015). Locomotor play drives motor skill acquisition at the expense of growth, a life history trade-off. *Science Advances*, 1, e1500451.

Bergman, T.J. & Beehner, J.C. (2008). A simple method for measuring colour in wild animals: validation and use on chest patch colour in geladas (*Theropithecus gelada*). *Biological Journal of the Linnean Society*, 94(2), 231–240.

Bergman, T.J. & Kitchen, D.M. (2009). Comparing responses to novel objects in wild baboons (*Papio ursinus*) and geladas (*Theropithecus gelada*). *Animal Cognition*, 12, 63–73.

Berkman, L.F. (1984). Assessing the physical health effects of social networks and social support. *Annual Review of Public Health*, 5, 413–432.

Berkman, L.F., Glass, T., Brissette, I. & Seeman, T.E. (2000). From social integration to health, Durkheim in the new millennium. *Social Science & Medicine*, 51, 843–857.

Bernstein, R.M., Leigh, S.R., Donovan, S.M. & Monaco, M.H. (2008). Hormonal correlates of ontogeny in baboons (*Papio hamadryas anubis*) and mangabeys (*Cercocebus atys*). *American Journal of Physical Anthropology*, 136, 156–168

Bibollet-Ruche, F., Galat-Luong, A., Cuny, G., Sarni-Manchado, P., Galat, G., Durand, J.P., Pourrut, X. & Veas, F. (1996). Simian immunodeficiency virus infection in a patas monkey (*Erythrocebus patas*), evidence for cross-species transmission from African green monkeys (*Cercopithecus aethiops sabaeus*) in the wild. *Journal of General Virology*, 77, 773–781.

Bielert, C., Czaja, J.A., Eisele, S., Scheffler, G., Robinson, J.A. & Goy, R.W. (1976). Mating in the rhesus monkey (*Macaca mulatta*) after conception and its relationship to oestradiol and progesterone levels throughout pregnancy. *Journal of Reproduction and Fertility*, 46, 179–187.

Birky, C.W., Maruyama, T., & Fuerst, P. (1983). An approach to population and evolutionary genetic theory for genes in mitochondria and chloroplasts, and some results. *Genetics*, 103(3), 513–527.

Blakeslee, J.R., Sowder, W.G. & Baulu, J. (1985). Wild African green monkeys of Barbados are HTLV negative. *Lancet*, 8427, 525.

Blanckenhorn, W.U. (2000). The evolution of body size, what keeps organisms small? *Quarterly Review of Biology*, 75 (4), 385–407.

Blaszczyk, M.B. (2017). Boldness towards novel objects predicts predator inspection in wild vervet monkeys. *Animal Behaviour*, 123, 91–100. https://doi.org/10.1016/j.anbehav.2016.10.017.

Blomquist, G.E. (2009). Environmental and genetic causes of maturational differences among rhesus macaque matrilines. *Behavioral Ecology & Sociobiology*, 63, 1345–1352.

Blomquist, G.E. & Turnquist, J.E. (2011). Selection on adult female body size in rhesus macaques. *Journal of Human Evolution*, 60, 677–683.

Bogin, B. (1999). *Patterns of Human Growth*, 2nd edn. Cambridge: Cambridge University Press.

Bolter, D.R. & Zihlman, A.L. (2003). Morphometric analysis of growth and development in wild-collected vervet monkeys (*Cercopithecus aethiops*), with implications for growth patterns across Old World monkeys, apes and humans. *Journal of Zoology*, 260(1), 99–110.

Bonduriansky, R. (2007a). Sexual selection and allometry, a critical reappraisal of the evidence and ideas. *Evolution*, 61, 838–849.

  (2007b). The evolution of condition-dependent sexual dimorphism. *American Naturalist*, 169, 9–19.

Bones and Behavior Working Group (2015). Bones and Behavior Working Group. Retrieved from www.bonesandbehavior.org

Bose, T., Voruganti, V.S., Tejero, M.E., Proffit, J.M., Cox, L.A., Vandeberg, J.L., Mahaney, M.C., Rogers, J., Freeland-Graves, J.H., Cole, S.A. & Comuzzie, A.G. (2010). Identification of a QTL for adipocyte volume and of shared genetic effects with aspartate aminotransferase. *Biochemical Genetics*, 48, 538–547.

Bosinger, S.E., Li, Q., Gordon, S.N., Klatt, N.R., Duan, L., Xu, L., Francella, N., Sidahmed, A., Smith, A.J., Cramer, E.M., Zeng, M., Masopust, D., Carlis, J.V., Ran, L., Vanderford, T.H., Paiardini, M., Isett, R.B., Baldwin, D.A., Else, J.G., Staprans, S.I., Silvestri, G., Haase, A.T. & Kelvin, D.J. (2009). Global genomic analysis reveals rapid control of a robust innate response in SIV-infected sooty mangabeys. *Journal of Clinical Investigation*, 119, 3556–3572.

Botero, S., Stevenson, P.R., & Di Fiore, A. (2015). A primer on the phylogeography of *Lagothrix lagotricha* (sensu Fooden) in northern South America. *Molecular Phylogenetics and Evolution*, 82, 511–517.

Boulton, A.M., Horrocks, J.A. & Baulu, J. (1995). Reevaluating the Barbados vervet monkey population and its role in crop damage. *American Journal of Primatology*, 36, 112.

Boulton, A.M., Horrocks, J.A. & Baulu, J. (1996). The Barbados vervet monkey, changes in population size and crop damage, 1980–1994. *International Journal of Primatology*, 17, 831–844.

Bradford, A.P., Jones, K., Kechris, K., Chosich, J., Montague, M., Warren, W.C., May, M.C., Al-Safi, A., Kuokkanen, S., Appt, S.E. & Polotsky, A.J. (2015). Joint miRNA/mRNA expression profiling reveals changes consistent with development of dysfunctional corpus luteum after weight gain. *PLoS One*, 10, e0135163.

Bradley, B.J. & Lawler, R.R. (2011). Linking genotypes, phenotypes, and fitness in wild primate populations. *Evolutionary Anthropology*, 20, 104–119.

Brain, C.K. (1965). Observations on the behavior of vervet monkeys (*Cercopithecus aethiops*). *Zoologica Africana*, 1, 13–27.

Brain, C. & Mitchell, D. (1999). Body temperature changes in free-ranging baboons (Papio hamadryas ursinus) in the Namib Desert, Namibia. *International Journal of Primatology*, 20, 585–598.

Bramblett, C.A., Bramblett, S.S., Bishop, D.A. & Coelho, A.M. (1982). Longitudinal stability in adult status hierarchies among vervet monkeys (*Cercopithecus aethiops*). *American Journal of Primatology*, 2, 43–51.

Brenchley, J.M. & Douek, D.C. (2012). Microbial translocation across the GI tract. *Annual Review of Immunology*, 30, 149–173.

Brenchley, J.M., Price, D.A. & Douek, D.C. (2006). HIV disease, fallout from a mucosal catastrophe? *Nature Immunology*, 7, 235–239.

Brenchley, J.M., Price, D.A., Schacker, T.W. et al. (2006). Microbial translocation is a cause of systemic immune activation in chronic HIV infection. *Nature Medicine*, 12, 1365–1371.

Breuer, T., Hockemba M.B.N., Olejniczak, C., Parnell, R.J. & Stokes, E.J. (2009). Physical maturation, life-history classes and age estimates of free-ranging western gorillas – insights from Mbeli Bai, Republic of Congo. *American Journal of Primatology*, 71(2), 106–119.

Bronikowski, A.M., Alberts, S.C., Altmann, J., Packer, C., Carey, K.D. & Tatar, M. (2002). The aging baboon: comparative demography in a non-human primate. *Proceedings of the National Academy of Sciences of the United States of America*, 99, 9591–9595.

Bronson, F.H. (1989). *Mammalian Reproductive Biology*. Chicago: University of Chicago Press.

Broussard, S.R., Staprans, S.I., White, R., Whitehead, E.M., Feinberg, M.B. & Allan, J.S. (2001). Simian immunodeficiency virus replicates to high levels in naturally infected African green monkeys without inducing immunologic or neurologic disease. *Journal of Virology*, 75, 2262–2275.

Brown, A.C. (2008). Status and range of introduced mammals on St. Martin, Lesser Antilles. *Living World, Journal of the Trinidad and Tobago Field Naturalist Club*, 14–18.

Buchan, J.C., Alberts, S.C., Silk, J.B. & Altmann, J. (2003). True paternal care in a multi-male primate society. *Nature*, 425(6954), 179–181.

Burness, G.P., Diamond, J. & Flannery, T. (2001). Dinosaurs, dragons and dwarfs, the evolution of maximum body size. *Proceedings of the National Academy of Sciences of the United States of America*, 98, 14518–14523.

Busse, C.D. (1977). Chimpanzee predation as a possible factor in the evolution of red colobus monkey social organization. *Evolution*, 31, 907–911.

Butynski, T.M. (1988). Guenon birth seasons and correlates with rainfall and food. In A. Gautier-Hion, F. Bourliere, J.P. Gautier-Hion & J. Kingdon, eds., *A Primate Radiation: Evolutionary Biology of the African Guenons.* Cambridge: Cambridge University Press, pp. 284–322.

Butynski, T.M., Kingdon, J. & Kalina, J. (2013). *Mammals of Africa: Primates.* London, UK: Bloomsbury.

Byles, R.H. & Sanders, M.F. (1981). Intertroop variation in the frequencies of ABO alleles in a population of olive baboons. *International Journal of Primatology*, 2, 35–46.

Byrnes, J.P., Miller, D.C. & Schafer, W.D. (1999). Gender differences in risk taking: a meta-analysis. *Psychological Bulletin*, 125(3), 367.

Caillaud, D., Levrero, F., Cristescu, R., Gatti, S., Dewas, M., Douadi, M., Gautier-Hion, A., Raymond, M. & Menard, N. (2006). Gorilla susceptibility to Ebola virus, the cost of sociality. *Current Biology*, 16, R489–R491.

Cameron, T.W. (1930). The species of *Subulura* Molin in primates. *Journal of Helminthology*, 8, 49–57.

Campbell-Smith, G., Simanjorang, H.V., Leader-Williams, N. & Linkie, M. (2010). Local attitudes and perceptions toward crop-raiding by orangutans (*Pongo abelii*) and other nonhuman primates in northern Sumatra, Indonesia. *American Journal of Primatology*, 72, 866–876.

Campos, F.A. & Fedigan, L.M. (2009). Behavioural adaptations to heat stress and water scarcity in white-faced capuchins (Cebus capucinus) in Santa Rosa National Park, Costa Rica. *American Journal of Physical Anthropology*, 138, 101–111.

Carbone, L.L., Harris, R.A., Gnarre, S. et al. (2014). Gibbon genome and the fast karyotype evolution of small apes. *Nature*, 513, 195–201.

Carmody, R.N. & Wrangham, R.W. (2009). The energetic significance of cooking. *Journal of Human Evolution*, 57(4), 379–391.

Caro, T.M. (2005). *Anti-Predator Defenses in Birds and Mammals.* Chicago: University of Chicago Press.

Caughley, G. (1966). Mortality patterns in mammals. *Ecology*, 47, 906–918.
(1977). *Analysis of Vertebrate Populations.* New York: John Wiley.

Cerling, T.E., Wittemyer, G., Rasmussen, H.B., Vollrath, F., Cerling, C.E., Robinson, T.J. & Douglas-Hamilton, I. (2006). Stable isotopes in elephant hair document migration patterns and diet changes. *Proceedings of the National Academy of the Sciences of United States of America*, 103, 371–373.

Chahroudi, A., Bosinger, S.E., Vanderford, T.H., Paiardini, M. & Silvestri, G. (2012). Natural SIV hosts, showing AIDS the door. *Science*, 335, 1188–1193.

Chakraborty, R. (1974). A note on Nei's measure of gene diversity in a substructured population. *Humangenetik*, 21(1), 85–88.

Chan, J.L. & Mantzoros, C.S. (2001). Leptin and the hypothalamtic–pituitary regulation of the gonadotropin–gonadal axis. *Pituitary*, 4(1–2), 87–92.

Chancellor, R.L. & Isbell, L.A. (2008). Punishment and competition over food in captive rhesus macaques (*Macaca mulatta*). *Animal Behaviour*, 75, 1939–1947.

(2009). Food site residence time and female competitive relationships in wild gray-cheeked mangabeys (*Lophocebus albigena*). *Behavioral Ecology and Sociobiology*, 63, 1447–1458.

Chapman, C.A. (1985). The influence of habitat on behavior in a group of St. Kitts green monkeys. *Journal of Zoology*, 206, 311–320.

(1987). Selection of secondary growth areas by vervet monkeys (*Cercopithecus aethiops*). *American Journal of Primatology*, 12, 217–221.

Chapman, C.A. & Balcomb, S.R. (1998). Population characteristics of howlers, ecological conditions or group history. *International Journal of Primatology*, 19(3), 385–403.

Chapman, C. & Fedigan, L.M. (1984). Territoriality in the St Kitts vervet, *Cercopithecus aethiops*. *Journal of Human Evolution*, 13(8), 677–686.

Chapman, C.A., Fedigan, L.M. & Fedigan, L. (1988). Ecological and demographic influences on the pattern of association in St. Kitts vervets. *Primates*, 29, 417–421.

Chapman, C.A., Wrangham, R.W. & Chapman, L.J. (1995). Ecological constraints on group size, an analysis of spider monkey and chimpanzee subgroups. *Behavioral Ecology and Sociobiology*, 36, 59–70.

Chapman, C.A., Wasserman, M.D., Gillespie, T.R., Speirs, M.L., Lawes, M.J., Saj, T.L. & Ziegler, T.E. (2006). Do food availability, parasitism, and stress have synergistic effects on red colobus populations living in forest fragments? *American Journal of Physical Anthropology*, 131(4), 525–534.

Chapman, C. A., Friant, S., Godfrey, K. et al. (2016). Social behaviours and networks of vervet monkeys are influenced by gastrointestinal parasites. *PLoS One*, 11(8), e0161113.

Chapman, C.A., Gillespie, T.R. & Speirs, M.L. (2005). Parasite prevalence and richness in sympatric colobines: Effects of host density. *American Journal of Primatology*, 67(2), 259–266.

Charnov, E.L. (1993). *Life History Invariants*. Oxford, UK: Oxford University Press.

Charpentier, M.J.E., Prugnolle, F., Giminez, P. & Widdig, A. (2008). Genetic heterozygosity and sociality in a primate species. *Behavioural Genetics*, 38(2), 151–158.

Charpentier, M.J.E., Tung, J., Altmann, J. & Alberts, S.C. (2008). Age at maturity in wild baboons: genetic, environmental and demographic influences. *Molecular Ecology*, 17(8), 2026–2040.

Charpentier, M.J., Fontaine, M.C., Cherel, E., Renoult, J.P., Jenkins, T., Benoit, L., Barthes, N., Alberts, S.C. & Tung, J. (2012). Genetic structure in a dynamic baboon hybrid zone corroborates behavioural observations in a hybrid population. *Molecular Ecology*, 21 (3), 715–731.

Chen, J.A., Fears, S.C., Jasinska, A.J. et al. (2018). Neurodegenerative disease biomarkers Aβ1–40, Aβ1–42, tau, and p-tau181 in the vervet monkey cerebrospinal fluid: Relation to normal aging, genetic influences, and cerebral amyloid angiopathy. *Brain and Behavior*, 8(2), e00903.

Cheney, D.L. (1981). Inter-group encounters among free-ranging vervet monkeys. *Folia Primatologica*, 35, 124–146.

(1983). Extrafamilial alliances among vervet monkeys. In R.A. Hinde, ed., *Primate Social Relationships: An Integrated Approach*. Sunderland, MA: Sineauer Associates, pp. 103–111.

(1987). Interactions and relationships between groups. In B.B. Smuts, D.L. Cheney R.M. Seyfarth, R.W. Wrangham & T.T. Struhsaker, eds., *Primate Societies*. Chicago: University of Chicago Press, pp. 267–281.

Cheney, D.L. & Seyfarth, R.M. (1980). Vocal recognition in free-ranging vervet monkey. *Animal Behavior*, 28, 362–367.

(1981). Selective forces affecting the predator alarm calls of vervet monkeys. *Behaviour*, 76, 25–61.

(1982). How vervet monkeys perceive their grunts: field playback experiments *Animal Behavior*, 30, 739–751.

(1983). Non-random dispersal in free-ranging vervet monkeys, social and genetic consequences. *The American Naturalist*, 122, 392–412.

(1986). The recognition of social alliances by vervet monkeys. *Animal Behaviour*, 34(6), 1722–1731.

(1987). The influence of intergroup competition on the survival and reproduction of female vervet monkeys. *Behavioral Ecology and Sociobiology*, 21(6), 375–386.

(1989). Redirected aggression and reconciliation among vervet monkeys, *Cercopithecus aethiops*. *Behaviour*, 110(1–4), 258–275.

(1990a). *How Monkeys See the World*. Chicago: University of Chicago Press.

(1990b). Attending to behaviour versus attending to knowledge: examining monkeys' attribution of mental states. *Animal Behaviour*, 40, 742–753.

(1998). Why monkeys don't have language. In G. Peterson, ed., *The Tanner Lectures on Human Values*. Salt Lake City, UT: University of Utah Press, p. 19.

Cheney, D.L. & Wrangham, R.W. (1987). Predation. In B.B. Smuts, D.L. Cheney R.M. Seyfarth, R.W. Wrangham & T.T. Struhsaker, eds., *Primate Societies*. Chicago: University of Chicago Press, pp. 227–239.

Cheney, D.L., Lee, P.C. & Seyfarth, R.M. (1981). Behavioral correlates of non-random mortality among free-ranging female vervet monkeys. *Behavioral Ecology and Sociobiology*, 9, 153–161.

Cheney, D.L., Seyfarth, R.M., Andelman, S.J. & Lee, P.A. (1988). Reproductive success in vervet monkeys. In T.H. Clutton-Brock, ed., *Reproductive Success: Studies of Individual Variation in Contrasting Breeding Systems.* Chicago: University of Chicago Press, pp. 384–402.

Cheverud, J.M., Wilson, P. & Dittus, W.P.J. (1992). Primate population studies at Polonnaruwa. II. Heritability of body measurements in a natural population of toque macaques (*Macaca sinica*). *American Journal of Primatology*, 27(2), 145–156.

Chiarelli, B. (1968). Chromosome polymorphism in the species of the genus *Cercopithecus. Cytologia*, 33, 1–16.

Chichester, L., Gee, M.K., Jorgensen, M.J. & Kaplan, J.R. (2015a). Hematology and clinical chemistry measures during and after pregnancy and age- and sex-specific reference intervals in African green monkeys (*Chlorocebus aethiops sabaeus*). *Journal of the American Association for Laboratory Animal Science*, 54, 359–367.

Chichester, L., Wylie, A.T., Craft, S. & Kavanagh, K. (2015b). Muscle heat shock protein 70 predicts insulin resistance with aging. *The Journals of Gerontology. Series A, Biological Sciences and Medical Sciences*, 70, 155–162.

Clarke, M.R. & Mayeaux, D.J. (1992). Aggressive and affiliative behavior in green monkeys with differing housing complexity. *Aggressive Behavior*, 18(3), 231–239.

Clarke, A.S., Kammerer, C.M., George, K.P., Kupfer, D.J., McKinney, W.T., Spence, M.A. & Kraemer, G.W. (1995). Evidence for heritability of biogenic amine levels in the cerebrospinal fluid of rhesus monkeys. *Biological Psychiatry*, 38, 572–577.

Clemente, J.C., Pehrsson, E.C., Blaser, M.J. et al. (2015). The microbiome of uncontacted Amerindians. *Science Advances*, 1(3), e1500183.

Cloninger, C.R., Przybeck, T.R., Svrakic, D.M. & Wetzel, R.D. (1994). *The Temperament and Character Inventory (TCI): A Guide to Its Development and Use.* St. Louis, MI: Center for the Psychobiology of Personality.

Clutton-Brock, T.H. & Harvey, P.H. (1977). Primate ecology and social organization. *Journal of Zoology*, 183, 1–39.

Clutton-Brock, T.H., Albon, S.D. & Guinness, F.E. (1985). Parental investment and sex differences in juvenile mortality in birds and mammals. *Nature*, 313, 131–133.

Clutton-Brock T.H, Guinness F.E. & Albon S.D. (1982). *Red Deer: Behaviour and Ecology of Two Sexes*. Chicago: University of Chicago Press.

Clutton-Brock, T.H., Harvey, P.H. & Rudder, B. (1977). Sexual dimorphism, socionomic sex ratio and body mass in primates. *Nature*, 269, 797–800.

Cockburn, A., Legge, S. & Double, M.C. (2002). Sex ratios in birds and mammals: can the hypotheses be disentangled? In I.C.W. Hardy, ed., *The Sex Ratio Handbook*. Cambridge: Cambridge University Press, pp. 266–286.

Coelho, A.M. & Rutenberg, G.W. (1989). Neonatal nutrition and longitudinal growth in baboons: adiposity measured by skinfold thickness. *American Journal of Human Biology*, 1(4), 429–442.

Coleman, B.T. & Hill, R.A. (2014). Living in a landscape of fear: the impact of predation, resource availability and habitat structure on primate range use. *Animal Behaviour*, 88, 165–173.

Collins, N., Dunkel-Schetter, C., Lobel, M. & Scrimshaw, S. (1993). Social support in pregnancy: psychosocial correlates of birth outcomes and postpartum depression. *Journal of Personality and Social Psychology*, 65, 1243–1258.

Compton, A.A. & Emerman, M. (2013). Convergence and divergence in the evolution of the APOBEC3G–Vif interaction reveal ancient origins of simian immunodeficiency viruses. *PLoS Pathogens*, 9, e1003135.

CONVERGE Consortium (2015). Sparse whole-genome sequencing identifies two loci for major depressive disorder. *Nature*, 523, 588–591.

Cormier, L.A. (2002). Monkey as food, monkey as child, Guaja symbolic cannibalism. In A. Fuentes & L.D. Wolfe., eds., *Primates Face to Face: The Conservation Implications of Human and Nonhuman Primate Interconnections*. Cambridge: Cambridge University Press, pp. 63–84.

(2003). *Kinship with Monkeys: The Guaja Foragers of Eastern Amazonia*. New York: Columbia University Press.

Coppenhaver, D. & Buettner-Janusch, J. (1970). Transferrins of Cercopithecinae. *Folia Primatologica*, 13, 23–34.

Coppenhaver, D. & Olivier, T.J. (1986). Immunoglobulin allotypes of Kenyan olive baboons: troop frequencies, linkage disequilibria, and comparisons with other studies. *International Journal of Primatology*, 7(4), 335–350.

Coppinger, R.P. & McGuire, J.P. (1980). *Cercopithecus aethiops* of St. Kitts, a population estimate based on human predation. *Caribbean Journal of Science*, 15, 1–7.

Cords, M. (1987). Mixed-species association in East African guenons: general patterns or a collection of specific examples? *International Journal of Primatology*, 8(5), 454.

(2000). The number of males in guenon groups. In P. Kappeler, ed., *Primate Males: Causes and Consequences of Variation in Group Composition*. Cambridge: Cambridge University Press, pp. 84–96.

Cortes-Ortiz, L., Bermingham, E., Rico, C., Rodriguez-Luna, E., Sampaio, I. & Ruiz-Garcia, M. (2003). Molecular systematics and biogeography of the Neotropical monkey genus, *Alouatta*. *Molecular Phylogenetics and Evolution*, 26(1), 64–81.

Costa, P.T., Terracciano, A. & McCrae, R.R. (2001). Gender differences in personality traits across cultures, robust and surprising findings. *Journal of Personality and Social Psychology*, 81, 322–331.

Cowlishaw, G. (1997). Trade-offs between foraging and predation risk determine habitat use in a desert baboon population. *Animal Behaviour*, 53, 667–686.

Cox, L.A., Comuzzie, A.G., Havill, L.M., Karere, G.M., Spradling, K.D., Mahaney, M.C., Nathanielsz, P.W., Nicolella, D.P., Shade, R.E., Voruganti, S. & VandeBerg, J.L. (2013). Baboons as a model to study genetics and epigenetics of human disease. *ILAR Journal*, 54, 106–121.

Colyer, F. (1948). Variations of the teeth of the green monkey in St. Kitts. *Journal of the Royal Society of Medicine*, 41, 845–848.

Craine, J.M., Elmore, A.J., Aidar, M.P.M., Bustamante, M., Dawson, T.E., Hobbie, E.A., Kahman, A., Mack, M., McLauchlan, K.K., Michelsen, A., Bardoto, G.B., Pardo, L.H., Penuela, J., Reich, P.B., Schuur, E.A.G., Stock, W.D, Templar, P.H., Virginia, R.S., Welker, J.M. & Wright, I.J. (2009). Global patterns of foliar nitrogen isotopes and their relationships with climate, mycorrhizal fungi, foliar nutrient concentrations, and nitrogen availability. *New Phytologist*, 183, 980–992.

Cramer, J.D., Gaetano, T.J., Gray, J.P., Grobler, J.P., Lorenz, J., Freimer, N., Schmitt, C.A. & Turner, T.R. (2013). Variation in scrotal color among widely distributed vervet monkey populations (*C. a. pygerythrus* and *C. a. sabaeus*). *American Journal of Primatology*, 75, 752–762.

Cramer, J.D., Gobler, J.P., Freimer, N., Turner, T.R. (2014). Blue sexual skin color and male dominance among South African vervet monkeys (*Ch. a. pygerythrus*). Paper presented at the biennial Congress of the International Primatological Society, Hanoi, Vietnam.

Cramer, A.E. & Gallistel, C.R. (1997). Vervet monkeys as travelling salesmen. *Nature*, 387, 464.

Crook, J.H. & Gartlan, J.S. (1966). Evolution of primate societies. *Nature*, 210, 1200–1203.

Cross, C.P., Cyrenne, D.M. & Brown, G.R. (2013). Sex differences in sensation-seeking, a meta-analysis. *Scientific Reports*, 3, 1–5.

Cuervo, J.J. & Møller, A.P. (2001). Components of phenotypic variation in avian ornamental and non-ornamental feathers. *Evolutionary Ecology*, 25, 53–72.

Cuozzo, F.P. & Sauther, M.L. (2006). Temporal change in tooth size among ringtailed lemurs (*Lemur catta*) at the Beza Mahafaly Special Reserve, Madagascar: effects of an environmental fluctuation. In A. Jolly, R.W. Sussman, N. Koyama & H. Rasamimanana, eds., *Ringtailed Lemur Biology, Lemur catta, in Madagascar*. New York: Springer, pp. 343–366.

Dall, S.R.X. & Griffith, S.C. (2014). An empiricist guide to animal personality variation in ecology and evolution. *Ecology & Evolution*, 2, 1–7.

Dandelot, P. (1959). Note sur la classification des Cercopitheques du groups aethiops. *Mammalia*, 23, 357–368.

(1971). Order Primates. In J. Meester & H. Setzer, eds., *The Mammals of Africa*. Washington, DC: Smithsonian Institution.

Danzy, J., Grobler, J.P., Freimer, N. & Turner, T.R. (2012). Sunbathing: a behavioral response to seasonal climatic change among South African vervet monkeys (*Chlorocebus aethiops*). *African Primates*, 7(2), 230–237.

Darga, L.L., Goodman, M., Weiss, M.L., Moore, G.W., Prychodko, W., Dene, H., Tashian, R. & Koen, A. (1975). Molecular systematics and clinal variation in macaques. In C.L. Markert, ed., *Isozymes, Volume 4, Genetics and Evolution. International Conference on Izozymes*. New York: Academic Press, pp. 797–812.

Darwin, C. (1859). *On the Origin of Species by Means of Natural Selection*. London, UK: Murray.

Darwin, C.R. (1874). *The Descent of Man, and Selection in Relation to Sex*, 2nd edn. New York: Appleton.

Dausmann, K.H., Glos, J., Ganzhorn, J.U. & Heldmaier, G. (2004). Physiology: hibernation in a tropical primate. *Nature*, 429, 825–826.

Davenport, M.D., Tiefenbacher, S., Lutz, C.K., Novak, M.A. & Meyer, J.S. (2006). Analysis of endogenous cortisol concentrations in the hair of rhesus macaques. *General and Comparative Endocrinology*, 147(3), 255–261.

de Jong, G., de Ruiter, J.R. & Harring, R. (1994). Genetic structure of a population with social structure and migration. In V. Loeschcke, J. Tomiuk & S.K. Jain, eds., *WXS 68, Conservation Genetics*. Basel, Switzerland: Birkhauser Verlag, pp. 147–164.

Delson, E. (1984). Cercopithecid biochronology of the African Plio-Pleistocene: correlation among eastern and southern hominid-bearing localities. *Courier Forschungsinstitut Senckenberg*, 69, 199–218.

(1992). Evolution of Old World monkeys. In S. Jones, R. Martin & D. Pilbeam, eds., *The Cambridge Encyclopedia of Human Evolution*. Cambridge: Cambridge University Press, pp. 217–222.

De Moor, P.P. & Steffens, F.E. (1972). The movements of vervet monkeys (*Cercopithecus aethiops*) within their ranges as revealed by radio-tracking. *Journal of Animal Ecology*, 41(3), 677–687.

Denham, W.W. (1971). Energy relations and some basic properties of primate social organization. *American Anthropologist*, 73, 77–95.

(1982a). History of green monkeys in the West Indies part I. Migration from Africa. *The Journal of the Barbados Museum and Historical Society*, 36, 211–228.

(1982b). History of green monkeys in the West Indies part II. Population dynamics of Barbadian monkeys. *The Journal of the Barbados Museum and Historical Society*, 36, 211–228.

(1987). West Indian green monkeys, problems in historical biogeography. In F.S. Szalay, ed., *Contributions to Primatology*, vol. 24. Basel, Switzerland: Karger, pp. 1–78.

Deputte, B.L. (1992). Life history of captive gray-cheeked mangabeys: physical and sexual development. *International Journal of Primatology*, 13(5), 509–531.

de Queiroz, K. (2005). Ernst Mayr and the modern concept of species. *Proceedings of the National Academy of Sciences of the United States of America*, 102, 6600–6607.

de Ruiter, J.R. (1992). Capturing wild long-tailed macaques (*Macaca fascicularis*). *Folia Primatologica*, 59(2), 89–104.

de Ruiter, J.R. & Geffen, E. (1998). Relatedness of matrilines, dispersing males and social groups in long-tailed macaques (*Macaca fascicularis*). *Proceedings, Biological Sciences, Royal Society*, 265(1391), 79–87.

Dethlefsen, L., Eckburg, P.B., Bik, E.M. & Relman, D.A. (2006). Assembly of the human intestinal microbiota. *Trends in Ecology and Evolution*, 21(9), 517–523.

DeVries, T. (2003). Animal and human behaviour, genes, and environment in comparative ethology of primates. *Revista de la Pontificia Universidad Catolica del Ecuador*, 71, 217–223.

Di Fiore, A. (2009). Genetic approaches to the study of dispersal and kinship in New World primates. In P.A. Garber, A. Estrada, J.C. Bicca-Marques, E.W. Heymann & K.B Strier, eds., *South American Primates, Comparative Perspectives in the Study of Behavior, Ecology, and Conservation*. New York: Springer, pp. 211–250.

Di Fiore, A., Link, A.L., Schmitt, C.A. & Spehar, S.N. (2009). Dispersal patterns in sympatric woolly and spider monkeys: integrating molecular and observational data. *Behaviour*, 146, 437–470.

Diop, O.M., Gueye, A., Dias-Tavares, M., Kornfeld, C., Faye, A., Ave, P., Huerre, M., Corbet, S., Barre-Sinoussi, F. & Muller-Trutwin, M.C. (2000). High levels of viral replication during primary simian immunodeficiency virus SIVagm infection are rapidly and strongly controlled in African green monkeys. *Journal of Virology*, 74, 7538–7547.

Disotell, T.R. (1996). The phylogeny of Old World monkeys. *Evolutionary Anthropology*, 5(1), 18–24.

(2000). Molecular systematic of the Cercopithecidae. In P.F. Whitehead & C.J. Jolly, eds., *Old World Monkeys*. Cambridge: Cambridge University Press, pp. 29–56.

Dixson, A.F. (1998). *Primate Sexuality: Comparative Studies of the Prosimians, Monkeys, Apes and Human Beings*. Oxford, UK: Oxford University Press.

Dobzhansky, T. (1937). Genetic nature of species differences. *The American Naturalist*, 71(735), 404–420.

Dominguez-Bello, M.G., Costello, E.K., Contreras, M. et al. (2010). Delivery mode shapes the acquisition and structure of the initial microbiota across multiple body habitats. *Proceedings of the National Academy of Science of the United States of America*, 107(26), 11971–11975.

Dore, K.M. (2013). An Anthropological Investigation of the Dynamic Human–Vervet Monkey Interface in St. Kitts, West Indies. PhD Dissertation. Milwaukee, WI: University of Wisconsin-Milwaukee.

(2017a). Ethnophoresy. *The International Encyclopedia of Primatology*, 1–7.

(2017b). Vervets in the Caribbean. *The International Encyclopedia of Primatology*, 1–3.

(2017c). Navigating the methodological landscape: ethnographic data expose the nuances of "the monkey problem" in St. Kitts, West Indies. In K.M. Dore, E.P. Riley & A. Fuentes, eds., *Ethnoprimatology: A Practical Guide to the Human–Nonhuman Primate Interface*. Cambridge: Cambridge University Press, pp. 219–231.

(2018). Ethnoprimatology without conservation: the political ecology of farmer–green monkey (Chlorocebus sabaeus) relations in St. Kitts, West Indies. *International Journal of Primatology*, 39 (5), 918–944.

Dore, K.M., Mill, A. & Gallagher, C. (2014). Preliminary report on the use of GPS/GSM tracking devices to estimate the vervet monkey (*Chlorocebus aethiops sabaeus*) population on the island of St. Kitts. *American Journal of Physical Anthropology*, 156(S60), 124.

Dore, K.M., Riley, E.P. & Fuentes, A. (2017). *Ethnoprimatology: A Practical Guide to the Human–Nonhuman Primate Interface*. Cambridge: Cambridge University Press.

Dore, K.M., Eller, A.R., & Eller, J.L. (2018). Identity construction and symbolic association in farmer–vervet monkey (*Chlorocebus aethiops sabaeus*) interconnections in St. Kitts. *Folia Primatologica*, 89, 63–80.

Dore, K.M., Sewell, D. Mattenet, E. & Turner, T.R. (in press). GIS and GPS techniques in an ethnoprimatological investigation of St. Kitts vervet monkey (*Chlorocebus aethiops sabaeus*) crop-raiding behavior. In C.A. Shaffer, F. Dolins & J.R. Hickey, eds., *GIS and GPS in Primatology: A Practical Guide to Spatial Analysis*. Cambridge: Cambridge University Press.

Douek, D.C., Betts, M.R., Hill, B.J., Little, S.J., Lempicki, R., Metcalf, J.A., Casazza, J., Yoder, C., Adelsberger, J.W., Stevens, R.A., Baseler, M.W., Keiser, P., Richman, D.D., Davey, R.T. & Koup, R.A. (2001). Evidence for increased T cell turnover and decreased thymic output in HIV infection. *Journal of Immunology*, 167, 6663–6668.

Douek, D.C., McFarland, R.D., Keiser, P.H., Gage, E.A., Massey, J.M., Haynes, B.F., Polis, M.A., Haase, A.T., Feinberg, M.B., Sullivan, J.L., Jamieson, B.D., Zack, J.A., Picker, L.J. & Koup, R.A. (1998). Changes in thymic function with age and during the treatment of HIV infection. *Nature*, 396, 690–695.

Douglas, M. (1966). *Purity and Danger: An Analysis of the Concepts of Pollution and Taboo*. London, UK: Routledge and Kegan Paul.

Downing, H.J., Benimadho, S., Bolstridge, M.C. & Klomfass, H.J. (1973). The ABO blood groups in vervet monkeys (*Cercopithecus pygerythrus* F. Cuvier). *Journal of Medical Primatology*, 2(5), 290–295.

Dracopoli, N.C. & Brett, F.L. (1982). Serum aminopeptidases in pregnant vervet monkeys (*Cercopithecus aethiops*). *Biochemical Genetics*, 20(9–10), 825–831.

Dracopoli, N.C., Brett, F.L., Turner, T.R. & Jolly, C.J. (1983). Patterns of genetic variability in the serum proteins of the Kenyan vervet monkey (*Cercopithecus aethiops*). *American Journal of Physical Anthropology*, 61(1), 39–49.

Dracopoli, N.C., Turner, T.R., Else, J.G., Jolly, C.J., Anthony, R., Gallo, R.C. & Saxinger, W.C. (1986). STLV-1 antibodies in feral populations of East African vervet monkeys (*Cercopithecus aethiops*). *International Journal of Cancer*, 38(4), 523–529.

Drea, C.M. (2015). D'scent of man: a comparative survey of primate chemosignaling in relation to sex. *Hormones & Behavior*, 68, 117–133.

Dubuc, C., Muniz, L., Hesitermann, M., Engelhardy, A. & Widdig, A. (2011). Testing the priority-of-access model in a seasonally breeding primate species. *Behavioral Ecology and Sociobiology*, 65, 1615–1627.

Duggelby, C. (1978). Blood group antigens and the population genetics of *Macaca mulatta* on Cayo Santiago I. Genetic differentiation of social groups. *American Journal of Physical Anthropology*, 48, 35–40.

Ducheminsky, N., Henzi, S.P. & Barrett, L. (2014). Responses of vervet monkeys in large troops to terrestrial and aerial predator alarm calls. *Behavioral Ecology*, 25, 1474–1484.

Dunbar, R.I.M. (1974). Observations on the ecology and social organization of the green monkey, *Cercopithecus sabaeus*, in Senegal. *Primates*, 15, 341–350.

(1991). Functional significance of social grooming in primates. *Folia Primatologica*, 57, 121–131.

(2010). The social role of touch in humans and primates: behavioural function and neurobiological mechanisms. *Neuroscience & Biobehavioral Reviews*, 34, 260–268.

Dunbar, R.I.M. & Dunbar, E.P. (1974). Ecological relations and niche separation between sympatric terrestrial primates in Ethiopia. *Folia Primatologica*, 21, 36–60.

Dunbar, R. & Spoors, M. (1995). Social networks, support cliques, and kinship. *Human Nature*, 6, 273–290.

Dunn, J.C., Halenar, L.B., Davies, T.G., Cristobal-Azkarate, J., Reby, D., Sykes, D., Dengg, S., Fitch, W.T. & Knapp, L.A. (2015). Evoluationary trade-off between vocal tract and testes dimensions in howler monkeys. *Current Biology*, 25, 2839–2844.

Durham, N.M. (1969). Sex differences in visual threat displays of West African vervets. *Primates*, 10, 91–95.

Dutrillaux, B. (1986). The evolutionary role of chromosomes: a new interpretation. *Annales de Genetique*, 29(2), 69–75.

Eberhard, W.G. (2009). Static allometry and animal genitalia. *Evolution*, 63, 48–66.

Eberhard, W.G., Huber, B.A., Rodríguez, R.L., Briceño, R.D., Salas, I. & Rodríguez, V. (1998). One size fits all? Relationships between the size and degree of variation in genitalia and other body parts in twenty species of insects and spiders. *Evolution*, 52, 415–431.

Eberhard, W.G., Rodríguez, R.L., Huber, B.A., Speck, B., Miller, H., Buzatto, B.A., & Machado, G. (2018). Sexual selection and static allometry: the importance of function. *Quarterly Review of Biology*, 93, 207–250.

Ebstein, R.P. (2006). The molecular genetic architecture of human personality, beyond self-report questionnaires. *Molecular Psychiatry*, 11, 427–45.

Ehleringer, J., Field, C., Lin, Z. & Kuo, C. (1986). Leaf carbon isotope and mineral composition in subtropical plants along an irradiance cline. *Oecologia*, 70, 520–526.

Eisenberg, J.F. (1981). *The Mammalian Radiations*. Chicago: University of Chicago Press.

Eley, R.M., Tarara, R.P., Worthman, C.M. & Else, J.G. (1989). Reproduction in the vervet monkey (*Cercopithecus aethiops*). III. The menstrual cycle. *American Journal of Primatology*, 17, 1–10.

Elliott, S.T., Wetzel, K.S., Francella, N., Bryan, S., Romero, D.C., Riddick, N.E., Shaheen, F., Vanderford, T., Derdeyn, C.A., Silvestri, G., Paiardini, M. & Collman, R.G. (2015). Dualtropic CXCR6/CCR5 simian immunodeficiency virus (SIV) infection of sooty mangabey primary lymphocytes, distinct coreceptor use in natural versus pathogenic hosts of SIV. *Journal of Virology*, 89, 9252–9261.

Emlen, S.T. & Oring, L.W. (1977). Ecology, sexual selection, and the evolution of mating systems. *Science*, 197(4300), 215–223.

Emlen, D.J. (2008). The evolution of animal weapons. *Annual Review of Ecology, Evolution, and Systematics*, 39, 387–413.

(2014). *Animal Weapons*. New York: Henry Holt.

Engel, G., Hungerford, L.L., Jones-Engel, L., Travis, D., Eberle, R., Fuentes, A., Grant, R., Kyes, R. & Schillaci, M. (2006). Risk assessment: a model for predicting cross-species transmission of simian foamy virus from macaques (*M. fascicularis*) to humans at a monkey temple in Bali, Indonesia. *American Journal of Primatology*, 28(9), 934–948.

Enstam, K.L. (2007). Effects of habitat structure on perceived risk of predation and anti-predator behavior of vervet (*Cercopithecus aethiops*) and patas (*Erthyrocebus patas*) monkeys. In S. Gursky & K.A.I Nekaris, eds., *Primate Anti-Predator Strategies*. New York: Springer Science and Business Media, pp. 306–336.

Erhart, E.M., Coehlo Jr., A.M. & Bramblett, C.A. (1997). Kin recognition by paternal half-siblings in captive *Papio cynocephalus*. *American Journal of Primatology*, 43(2), 147–157.

Essock-Vitale, S. & Seyfarth, R.M. (1987). Intelligence and social cognition. In B.B. Smuts, D.L. Cheney, R.M. Seyfarth, R.W. Wrangham & T.T. Struhsaker, eds., *Primate Societies*. Chicago: University of Chicago Press, pp. 452–461.

Estes, J.D., Gordon, S.N., Zeng, M., Chahroudi, A.M., Dunham, R.M., Staprans, S.I., Reilly, C.S., Silvestri, G. & Haase, A.T. (2008). Early resolution of acute immune activation and induction of PD-1 in SIV-infected sooty mangabeys distinguishes nonpathogenic from pathogenic infection in rhesus macaques. *Journal of Immunology*, 180, 6798–6807.

Estrada, A., Garber, P.A., Rylands, A.B., Roos, C., Fernandez-Duque, E., Di Fiore, A., Nekaris, K.A.I., Nijman, V., Heymann, E.W., Lambert, J.E. & Rovero, F. (2017). Impending extinction crisis of the world's primates: why primates matter. *Science Advances*, 3, e1600946.

Fainman, J., Eid, M.D., Ervin, F.R. & Palmour, R.M. (2007). A primate model for Alzheimer's disease, investigation of the apolipoprotein E profile of the vervet monkey of St. Kitts. *American Journal of Medical Genetics. Part B, Neuropsychiatric Genetics*, 144B, 818–819.

Fairbanks, L.A. (1990). Reciprocal benefits of allomothering for female vervet monkeys. *Animal Behaviour*, 40(3), 553–562.

(2001). Individual differences in response to a stranger, social impulsivity as a dimension of temperament in vervet monkeys (*Cercopithecus aethiops sabaeus*). *Journal of Comparative Psychology*, 115, 22–28.

Fairbanks, L.A. & Bird, J. (1978). Ecological correlates of interindividual distance in the St. Kitts vervet (*Cercopithecus aethiops sabaeus*). *Primates*, 19, 605–614.

Fairbanks, L.A. & Hinde, K. (2013). Behavioral response of mothers and infants to variation in maternal condition: adaptation, compensation, and resilience. In K.B.H. Clancy, K. Hinde & J.N. Rutherford, eds., *Building Babies, Primate Development in Proximate and Ultimate Perspective*. New York: Springer, pp. 281–302.

Fairbanks, L.A. & Jorgensen, M.J. (2011). Objective behavioral tests of temperament in nonhuman primates. In A. Weiss. J.E. King & L. Murray, eds., *Personality and Temperament in Nonhuman Primates*. New York: Springer, pp. 103–128.

Fairbanks, L.A. & McGuire, M.T. (1984). Determinants of fecundity and reproductive success in captive vervet monkeys. *American Journal of Primatology*, 7(1), 27–38.

(1985). Relationships of vervet mothers with sons and daughters from one through three years of age. *Animal Behaviour*, 33(1), 40–50.

(1986). Age, reproductive value, and dominance-related behaviour in vervet monkey females: cross-generational influences on social relationships and reproduction. *Animal Behaviour*, 34(6), 1710–1721.

(1987). Mother–infant relationships in vervet monkeys, response to new adult males. *International Journal of Primatology*, 8, 351–366.

(1988). Long-term effects of early mothering behavior on responsiveness to the environment in vervet monkeys. *Developmental Psychobiology*, 21, 711–724.

Fairbanks L.A. & McGuire, M.T. (1993). Maternal protectiveness and response to the unfamiliar in vervet monkeys. *American Journal of Primatology*, 30, 119–129.

Fairbanks, L.A., Bailey, J.N., Breidenthal, S.E., Laudenslager, M.L., Kaplan, J.R. & Jorgensen, M.J. (2011). Environmental stress alters genetic regulation of novelty seeking in vervet monkeys. *Genes, Brain, and Behavior*, 10, 683–688.

Fairbanks, L.A., Fontenot, M.B., Phillips-Conroy, J.E., Jolly, C.J., Kaplan, J.R. & Mann, J.J. (1999). CSF monoamines, age and impulsivity in wild grivet monkeys (*Cercopithecus aethiops aethiops*). *Brain, Behavior and Evolution*, 53(5–6), 305–312.

Fairbanks, L.A., Jorgensen, M.J., Huff, A., Blau, K., Hung, Y.Y. & Mann, J.J. (2004a). Adolescent impulsivity predicts adult dominance attainment in male vervet monkeys. *American Journal of Primatology*, 64, 1–17.

Fairbanks, L.A., Jorgensen, M.J., Bailey, J.N., Breidenthal, S.E., Grzywa, R. & Laudenslager, M.L. (2011). Heritability and genetic corrclation of hair cortisol in vervet monkeys in low and higher stress environments. *Psychoneuroendocrinology*, 36(8), 1201–1208.

Fairbanks, L.A., Melega, W.P., Jorgensen, M.J., Kaplan, J.R. & McGuire, M.T. (2001). Social impulsivity inversely associated with CSF 5-HIAA and fluoxetine exposure in vervet monkeys. *Neuropsychopharmacology*, 24, 370–378.

Fairbanks, L.A., Newman, T.K., Bailey, J.N., Jorgensen, M.J., Breidenthal, S.E., Ophoff, R.A., Comuzzie, A.G., Martin, L.J. & Rogers, J. (2004b). Genetic contributions to social impulsivity and aggressiveness in vervet monkeys. *Biological Psychiatry*, 55, 642–647.

Fairbanks, L.A., Way, B.M., Breidenthal, S.E., Bailey, J.N. & Jorgensen, M.J. (2012). Maternal and offspring dopamine D4 receptor genotypes interact to influence juvenile impulsivity in vervet monkeys. *Psychological Science*, 23, 1099–1104.

Faith, J.J., McNulty, N.P., Rey, F.E. & Gordon, J.I. (2011). Predicting a human gut microbiota's response to diet in gnotobiotic mice. *Science*, 333, 101–104.

Faraone, S.V., Doyle, A.E., Mick, E. & Biederman, J. (2001). Meta-analysis of the association between the 7-repeat allele of the dopamine D_4 receptor gene and attention deficit hyperactivity disorder. *American Journal of Psychiatry*, 158, 1052–1057.

Faucheux, B., Bertraud, M. & Bourliere, F. (1978). Some effects of living conditions upon the pattern of growth in the stumptailed macaque (*Macaca arctoides*). *Folia Primatologica*, 30, 220–236.

Fedigan, L. & Fedigan, L. (1988). *Cercopithecus aethiops*: a review of field studies. In A. Gautier-Hion, F. Bourliere, J.P. Gautier-Hion & J. Kingdon, eds., *A Primate Radiation: Evolutionary Biology of the African Guenons*. Cambridge: Cambridge University Press, pp. 389–411.

Fedigan, L.M., Fedigan, L., Chapman, C. & McGuire, M.T. (1984). A demographic model of colonization by a population of St. Kitts vervets. *Folia Primatologica*, 42(3–4), 194–202.

Finch, C.E. & Rose, M.R. (1995). Hormones and the physiological architecture of life history evolution. *Quarterly Review of Biology*, 70(1), 1–52.

Fish, K.D., Sauther, M.L., Loudon, J.E. & Cuozzo, F.P. (2007). Coprophagy by wild ring-tailed lemurs (*Lemur catta*) in human-disturbed locations adjacent to the Beza Mahafaly Special Reserve, Madagascar. *American Journal of Primatology*, 69(6), 713–718.

Fooden, J. & Izor, R.J. (1983). Growth curves, dental emergence norms, and supplementary morphological observations in known-age captive orangutans. *American Journal of Primatology*, 5(4), 285–301.

Fooden, J. & Albrecht, G.H. (1993). Latitudinal and insular variation of skull size in crab-eating macaques (Primates, Cercopithecidae, *Macaca fascicularis*). *American Journal of Physical Anthropology*, 92(4), 521–538.

Fournier, D.A., Skaug, H.J., Ancheta, J. et al. (2012). AD Model Builder: using automatic differentiation for statistical inference of highly parameterized complex nonlinear models. *Optimization Methods and Software*, 27(2), 233–249.

Franceschini, M., Ziegler, T.E., Scheffler, G., Kaufman, G.E. & Sollod, A. (1997). A comparative analysis of fecal cortisol concentrations between four populations of woolly monkeys (*Lagothrix lagotricha*) living under different environmental conditions. *American Association of Zoo Veterinarians Annual Conference Proceedings*, 1997, 303–305.

Freeman, A.S., Kinsella, J.M., Cipolletta, C., Deem, S.L. & Karesh, W.B. (2004) Endoparasites of western lowland gorillas (*Gorilla gorilla gorilla*) at Bai Hokou, Central African Republic. *Journal of Wildlife Diseases*, 40(4), 775–781.

Freimer, N.B. Service, S.K., Ophoff, R.A., Jasinska, A.J., McKee, K., Villeneuve, A., Belisle, A., Bailey, J.N., Breidenthal, S.E., Jorgensen, M.J., Mann, J.J., Cantor, R.M., Dewar, K. & Fairbanks, L.A. (2007). A quantitative trait locus for variation in dopamine metabolism mapped in a primate model using reference sequences from related species. *Proceedings of the National Academy of Sciences of the United States of America*, 104, 15811–15816.

Frisancho, A.R. (1978). Human growth and development among high-altitude populations. In P.T. Baker, ed., *The Biology of High Altitutude Peoples*. Cambridge, UK: Cambridge University Press, pp. 117–171.

    & Baker, P.T. (1970). Altitude and growth: a study of the patterns of physical growth of a high altitude Peruvian Quechua population. *American Journal of Physical Anthropology*, 32(2), 279–292.

Frost, S.R. & Delson, E. (2002). Fossil Cercopithecidae from the Hadar Formation and surrounding areas of the Afar Depression, Ethiopia. *Journal of Human Evolution*, 43(5), 687–748.

    & Kullmer, O. (2008). Cercopithecidae from the Pliocene Chiwondo beds, Malawi-rift. *Geobios*, 41(6), 743–749.

Frost, S.R., Jablonski, N.G. & Haile-Selassie, Y. (2014). Early Pliocene Cercopithecidae from Woranso-Mille (Central Afar, Ethiopia) and the origins of the *Theropithecus oswaldi* lineage. *Journal of Human Evolution*, 76, 39–53.

Fruteau, C., Voelkl, B., Van Damme, E. & Noë, R. (2009). Supply and demand determine the market value of food providers in wild vervet monkeys. *Proceedings of the National Academy of Sciences*, 106(29), 12007–12012.

Fuentes, A. (2006). Human culture and monkey behavior: assessing the contexts of potential pathogen transmission between macaques and humans. *American Journal of Primatology*, 68(9), 880–896.

    (2010). Naturecultural encounters in Bali: monkeys, temples, tourists, and ethnoprimatology. *Cultural Anthropology*, 25, 600–624.

(2012). Ethnoprimatology and the anthropology of the human–primate interface. *Annual Review of Anthropology*, 41, 101–117.

(2014). Social minds and social selves: the human–alloprimate interface. In R. Corbey & A. Lanjouw, eds., *The Politics of Species: Reshaping Our Relationships with Other Animals*. Cambridge: Cambridge University Press, pp. 179–188.

Fuentes, A. & Hockings, K.J. (2010). The ethnoprimatological approach in primatology. *American Journal of Primatology*, 72(10), 841–847.

Fuentes, A., Southern, M. & Suaryana, K.G. (2005). Monkey forests and human landscapes: Is extensive sympatry sustainable for *Homo sapiens* and *Macaca fascicularis* in Bali? In J.D. Patterson & J. Wallis, eds., *Commensalism and Conflict: The Primate–Human Interface*. Norman, OK: American Society of Primatology Publications, pp. 168–195.

Fuentes, A. & Wolf, L. (2002). *Primates Face to Face: Conservation Implications of Human–Nonhuman Primate Interconnections*. New York: Cambridge University Press.

Fukasawa, M., Miura, T., Hasegawa, A., Morikawa, S., Tsujimoto, H., Miki, K., Kitamura, T. & Hayami, M. (1988). Sequence of simian immunodeficiency virus from African green monkey, a new member of the HIV/SIV group. *Nature*, 333, 457–461.

Fuller, A., Mitchell, D., Maloney, S.K. & Hetem, R.S. (2016). Towards a mechanistic understanding of the responses of large terrestrial mammals to heat and aridity associated with climate change. *Climate Change Responses*, 3, 10.

Gaetano, T.J., Danzy Cramer, J., Mtshali, M.S., Theron, N., Schmitt, C.A., Grobler, J.P., Freimer, N. & Turner, T.R. (2014). Mapping correlates of parasitism in wild South African vervet monkeys (*Chlorocebus aethiops*). *South African Journal of Wildlife Research*, 44(1), 56–70.

Gage, T.B. (1998). The comparative demography of primates: with some comments on the evolution of life histories. *Annual Review of Anthropology*, 27, 197–221.

Gagneux, P., Gonder, M.K., Goldberg, T.L. & Morin, P.A. (2001). Gene flow in wild chimpanzee populations: what genetic data tells us about chimpanzee movement over space and time. *Philosophical Transactions of the Royal Society of London B*, 356(1410), 889–897.

Galbany, J., Stoinski, T.S., Abavandimwe, D., Breuer, T., Rutkowski, W., Batista, N.V., Ndagijimana, F. & McFarlin, S.C. (2016). Validation of two independent photogrammetric techniques for determining body measurements of gorillas. *American Journal of Primatology*, 78, 418–431.

Galef, B.G. (2012). Social learning and traditions in animals: evidence, definitions, and relationship to human culture. *Wiley Interdisciplinary Reviews – Cognitive Science*, 3, 581–592.

Gao, F., Bailes, E., Robertson, D.L., Chen, Y., Rodenburg, C.M., Michael, S.F., Cummins, L.B., Arthur, L.O., Peeters, M., Shaw, G.M., Sharp, P.M. & Hahn, B.H. (1999). Origin of HIV-1 in the chimpanzee *Pan troglodytes troglodytes*. *Nature*, 397, 436–441.

Gartlan, J.S. (1969). Sexual and maternal behavior of the vervet monkey, *Cercopithecus aethiops. Journal of Reproduction and Fertility, Supplement*, 6(1), 137–150.

Gartlan, J.S. & Brain, C.K. (1968). Ecology and social variability in *Cercopithecus aethiops* and *C. mitis*. In P.C. Jay, ed., *Primates: Studies in Adaptation and Variability*. New York: Rinehart and Winston, pp. 253–292.

Gaulin, S.J.C. (1978). Activity, diet and dietary choice in an Andean population of *Alouatta seniculus. American Journal of Physical Anthropology*, 48, 397.

Gaulin, J.C. & Konner, M. (1977). On the natural diet of primates, including humans. *Nutrition and the Brain*, 1, 1–86.

Gautier, J.P. (1988). Interspecific affinities among guenons as deduced from vocalizations. In A. Gautier-Hion, F. Bourliere, J.P. Gautier-Hion & J. Kingdon, eds., *A Primate Radiation: Evolutionary Biology of the African Guenons*. Cambridge: Cambridge University Press, pp. 452–476.

Gautier, J.P. & Gautier-Hion, A. (1968). Polyspecific associations among the Cercopithecidae in Gabon. *Terre et la Vie*, 23, 164–201.

Gautier-Hion, A., Bourliére, F. & Gautier, J.P. (1988). *A Primate Radiation: Evolutionary Biology of the African Guenons*. Cambridge: Cambridge University Press.

Gavan, J.A. & Swindler, D.R. (1966). Growth rates and phylogeny in primates. *American Journal of Physical Anthropology*, 24, 181–190.

Gebo, D.L. & Sargis, E.J. (1994). Terrestrial adaptations in the postcranial skeletons of guenons. *American Journal of Physical Anthropology*, 93(3), 341–371.

Gemmill, A. & Gould, L. (2008). Microhabitat variation and its effects on dietary composition and intragroup feeding interactions between adult female *Lemur catta* during the dry season at Beza Mahafaly Special Reserve, southwestern Madagascar. *International Journal of Primatology*, 29, 1511–1533.

Genoud, M. (2002). Comparative studies of basal rate of metabolism in primates. *Evolutionary Anthropology*, 11(S1), 108–111.

Gerald, M.S. (1999). Scrotal Color in Vervet Monkeys (Cercopithecus aethiops sabaeus): The Signal Functions and Potential Proximate Mechanisms of Color Variation. PhD dissertation. Los Angeles, CA: University of California, Los Angeles.

(2001). Primate colour predicts social status and aggressive outcome. *Animal Behaviour*, 61, 559–566.

Gerald, M.S., Ayala, J., Ruiz-Lambides, A., Waitt, C. & Weiss, A. (2010). Do females pay attention to secondary sexual coloration in vervet monkeys (*Chlorocebus aethiops*)? *Naturwissenschaften*, 97(1), 89–96.

Gerloff, U., Hartung, B., Fruth, B., Hohmann, G. & Tautz, D. (1999). Intracommunity relationships, dispersal pattern and paternity success in a wild living community of bonobus (*Pan paniscus*) determined from DNA analysis of faecal samples. *Proceedings of the Royal Society of London B*, 266(1424), 1189–1195.

German, R.Z. (2004). The ontogeny of sexual dimorphism, the implications of longitudinal vs. cross-sectional data for studying heterochrony in mammals. In F. Anapol & N.G. Jablonski, eds., *Shaping Primate Evolution*. Cambridge: Cambridge University Press, pp. 11–23.

Gilbert, C.C., Frost, S.R. & Delson, E. (2016). Reassessment of Olduvai Bed I cercopithecoids: a new biochronological and biogeographical link to the South African fossil record. *Journal of Human Evolution*, 92, 50–59.

Gillespie, T.R. & Chapman, C.A. (2006). Prediction of parasite infection dynamics in primate metapopulations based on attributes of forest fragmentation. *Conservation Biology*, 20(2), 441–448.

Gillespie, T.R., Greiner, E.C. & Chapman, C.A. (2004). Gastrointestinal parasites of the guenons of western Uganda. *Journal of Parasitology*, 90(6), 1356–1360.

Gillespie, T.R, Greiner, E.C. & Chapman, C.A. (2005). Gastrointestinal parasites of the colobus monkeys of Uganda. *Journal of Parasitology*, 91(3), 569–573.

Glassman, D.M., Coelho Jr., A.M., Carey, K.D. & Bramblett, C.A. (1984). Weight growth in savannah baboons: a longitudinal study from birth to adulthood. *Growth*, 48(4), 425–433.

Glenn, M.E. & Cords, M. (2002). *The Guenons: Diversity and Adaptation in African Monkeys* (Vol. 2). New York: Springer Science & Business Media.

Gokcumen, O., Tischler, V. & Tica, J. (2013). Primate genome architecture influences structural variation mechanisms and functional consequences. *Proceedings of the National Academy of Sciences of the United States of America*, 110, 15764–15769.

Goldman, E.N. & Loy, J. (1997). Longitudinal study of dominance relations among captive patas monkeys. *American Journal of Primatology*, 42, 41–51.

Goldberg, T.L. & Wrangham, R.W. (1997). Genetic correlates of social behavior in wild chimpanzees: evidence from mitochondrial DNA. *Animal Behaviour*, 54(3), 559–570.

Goldstein, S., Brown, C.R., Ourmanov, I., Pandrea, I., Buckler-White, A., Erb, C., Nandi, J.S., Foster, G.J., Autissier, P., Schmitz, J.E. & Hirsch, V.M. (2006). Comparison of simian immunodeficiency virus SIVagmVer replication and

CD4$^+$ T-cell dynamics in vervet and sabaeus African green monkeys. *Journal of Virology*, 80, 4868–4877.

Goldstein, S., Ourmanov, I., Brown, C.R., Beer, B.E., Elkins, W.R., Plishka, R., Buckler-White, A. & Hirsch, V.M. (2000). Wide range of viral load in healthy African green monkeys naturally infected with simian immunodeficiency virus. *Journal of Virology*, 74, 11744–11753.

Gommery, D., Thackeray, J.F., Sénégas, F., Potze, S. & Kgasi, L. (2008). The earliest primate (*Parapapio* sp.) from the Cradle of Humankind World Heritage site (Waypoint 160, Bolt's Farm, South Africa). *South African Journal of Science*, 104(9–10), 405–408.

Goodman, M. & Poulik, E. (1961). Serum transferrins in the genus *Macaca*: species distribution of nineteen phenotypes. *Nature*, 191, 1407–1408.

Goodman, M., Kulkarni, A., Poulik, E. & Reklys, E. (1965). Species and geographic differences in the transferrin polymorphism of macaques. *Science*, 147, 884–886.

Gordon, A.D. (2006). Scaling of size and dimorphism in primates II: macroevolution. *International Journal of Primatology*, 27, 63–105.

Gordon, D., Huddleston, J. & Chaisson, M.J. (2016). Long-read sequence assembly of the gorilla genome. *Science*, 352, aae0344.

Gordon, S., Klatt, N.R., Bosinger, S.E., Brenchley, J.M., Milush, J.M., Engram, J.C., Dunham, R.M., Paiardini, M., Klucking, S., Danesh, A., Strobert, E.A., Apetrei, C., Pandrea, I.V., Kelvin, D., Douek, D.C., Staprans, S.I., Sodora, D.L. & Silvestri, G. (2007). Severe depletion of mucosal CD4$^+$ T cells in AIDS-free SIV-infected sooty mangabeys. *Journal of Immunology*, 179, 3026–3034.

Grobler, J.P. & Turner, T.R. (2010). A novel trap design for the collection and sedation of vervet monkeys (*Chlorocebus aethiops*). *South African Journal of Wildlife Research*, 40, 163–168.

Grobler, P., Jacquier, M., deNys, H., Blair, M., Whitten, P. L. & Turner, T. R. (2006). Primate sanctuaries, taxonomy and survival: a case study from South Africa. *Ecological and Environmental Anthropology (University of Georgia)*, 2.

Grobler, J.P., Coetzer, G., Dore, K., Lorenz, J., Schmitt, C.A., Freimer, N. & Turner, T.R. (2012). Genetic differentiation in populations of vervet monkeys (*Chlorocebus aethiops*) in South Africa. Paper presented at the Congress of the International Primatological Society, Cancun, Mexico.

Groman, S.M., James, A.S. & Seu, E. (2014). In the blink of an eye, relating positive-feedback sensitivity to striatal dopamine D$_2$-like receptors through blink rate. *Journal of Neuroscience*, 34, 14443–14454.

Grossman, Z., Meier-Schellersheim, M., Paul, W.E. & Picker, L.J. (2006). Pathogenesis of HIV infection: what the virus spares is as important as what it destroys. *Nature Medicine*, 12, 289–295.

Groves, C.P. (1989). *A Theory of Human and Primate Evolution*. Oxford, UK: Oxford Clarendon Press.

(2000). The phylogeny of the Ceropithecoidea. In P.F. Whitehead & C.J. Jolly, eds., *Old World Monkeys*. Cambridge, UK: Cambridge University Press, pp. 77–98.

(2001). *Primate Taxonomy*. Washington, DC: Smithsonian Institution Press.

(2005). Order Primates. In D.E. Wilson & D.M. Reeder, eds., *Mammal Species of the World*. Baltimore, MD: The Johns Hopkins University Press, pp. 111–184.

Grubb, P. (2006). Geospecies and superspecies in the African primate fauna. *Primate Conservation*, 20, 75–78.

Grubb, P., Butynski, T.M., Oates, J.F., Beader, S.K., Disotell, T.R., Groves, C.P. & Struhsaker, T.T. (2003). Assessment of the diversity of African primates. *International Journal of Primatology*, 24(6), 1301–1357.

Guschanski, K., Krause, J., Sawyer, S., Valenta, L.M., Bailey, S., Finstermeier, K., Sabin , R., Gilissen, E., Sonet, G., Nagy, Z.T., Georges, L., Mayer, F. & Savolainen, V. (2013). Next-generation museomics disentangles one of the largest primate radiations. *Systems Biology*, 62(4), 539–554.

Guy, A.J., Curnoe, D. & Banks, P.B. (2013). A survey of current mammal rehabilitation and release practices. *Biodiversity and Conservation*, 22(4), 825–837.

Guy, A.J., Stone, O.M.L. & Curnoe, D. (2011). The release of a troop of rehabilitated vervet monkeys (*Chlorocebus aethiops*) in KwaZulu-Natal, South Africa: outcomes and assessment. *Folia Primatologica*, 82(6), 308–320.

(2012). Assessment of the release of rehabilitated vervet monkeys into the Ntendeka Wilderness Area, KwaZulu-Natal, South Africa: a case study. *Primates*, 53(2), 171–179.

Hahn, N.E., Proulx, D., Muruthi, P.M., Alberts, S. & Altmann, J. (2003). Gastrointestinal parasites in free-ranging Kenyan baboons (*Papio cynocephalus* and *P. anubis*). *International Journal of Primatology*, 24(2), 271–279.

Hahn, B.H. (1994). Infection of yellow baboon with simian immunodeficiency virus from African green monkeys: evidence for cross-species transmission in the wild. *Journal of Virology*, 68(12), 8454–8460.

Hales, C.N. & Barker, D.J. (2001). The thrifty phenotype hypothesis. *British Medical Bulletin*, 60, 5–20.

Hamada, Y., Iwamoto, M. & Watanabe, T. (1986). Somatometrical features of Japanese monkeys in the Koshima Islet: in viewpoint of somatometry, growth, and sexual maturation. *Primates*, 27(4), 471–484.

Hamilton, W.D. (1971). Geometry for the selfish herd. *Journal of Theoretical Biology*, 31, 295–311.

Hannah, A.C. & McGrew, W.C. (1991). Rehabilitation of captive chimpanzees. In H.O. Box, ed., *Primate Responses to Environmental Change*. London, UK: Chapman & Hall, pp. 167–186.

Hanya, G. (2009). Effects of food type and number of feeding sites in a tree on aggression during feeding in wild *Macaca fuscata*. *International Journal of Primatology*, 30, 569–581.

Harcourt, A.H. (1987). Dominance and fertility among female primates. *Journal of Zoology*, 213, 471–487.

Hardin, R. & Remis, M.J. (2006). Biological and cultural anthropology of a changing tropical forest, a fruitful collaboration across subfields. *American Anthropologist*, 108, 273–285.

Harris, L.D., Tabb, B., Sodora, D.L., Paiardini, M., Klatt, N.R., Douek, D.C., Silvestri, G., Muller-Trutwin, M., Vasile-Pandrea, I., Apetrei, C., Hirsch, V., Lifson, J., Brenchley, J.M. & Estes, J.D. (2010). Downregulation of robust acute type I interferon responses distinguishes nonpathogenic simian immunodeficiency virus (SIV) infection of natural hosts from pathogenic SIV infection of rhesus macaques. *Journal of Virology*, 84, 7886–7891.

Harrison, G.A., Tanner, J.M., Pilbeam, D.R. & Baker, P.T. (1988). *Human Biology: An Introduction to Human Evolution, Variation, Growth, and Adaptibility*, 3rd edn. Oxford, UK: Oxford University Press.

Harrison, M.J.S. (1983) Age and sex differences in the diet and feeding strategies of the green monkey, *Cercopithecus sabaeus*. *Animal Behaviour*, 31(4), 969–977.

Harrison, T. (1989). New postcranial remains of *Victoriapithecus* from the middle Miocene of Kenya. *Journal of Human Evolution*, 18(1), 3–54.

Haus, T., Akom, E., Agwanda, B., Hofreiter, M., Roos, C. & Zinner, D. (2013). Mitochondrial diversity and distribution of African green monkeys (*Chlorocebus gray*, 1870). *American Journal of Primatology*, 75, 350–360.

Haus, T., Ferguson, B., Rogers, J., Doxiadis, G., Certa, U., Rose, N.J., Teepe, R., Weinbauer, G.F. & Roos, C. (2014). Genome typing of nonhuman primate models: implications for biomedical research. *Trends in Genetics*, 30, 482–487.

Hausfater, G. & Meade, B.J. (1982). Alternation of sleeping groves by yellow baboons (*Papio cynocephalus*) as a strategy for parasite avoidance. *Primates*, 23(2), 287–297.

Harwich, M.D., Serrano, M.G., Fettweis, J.M., Alves, J.M.P., Reimers, M.A., Buck, G.A. & Jefferson, K.K. (2012). Genomic sequence analysis and characterization of *Sneathia amnii* sp. nov. *BMC Genomics*, 13, S4.

Hawkes, K., Smith, K.R. & Robson, S.L. (2009). Mortality and fertility rates in humans and chimpanzees, how within-species variation complicates cross-species comparisons. *American Journal of Human Biology*, 21, 578–586.

Hector, A., Seyfarth, R. & Raleigh, M.J. (1989). Male parental care, female mate choice, and the effect of an audience in vervet monkeys. *Animal Behavior*, 38(2), 262–271.

Hellerstein, M.K., Hoh, R.A., Hanley, M.B. et al. (2003). Subpopulations of long-lived and short-lived T cells in advanced HIV-1 infection. *Journal of Clinical Investigation*, 112, 956–966.

Henzi, S.P. (1985). Genital signaling and the coexistence of male vervet monkeys (*Cercopithecus aethiops pygerythrus*). *Folia Primatologica*, 45, 129–147.

Henzi, S.P. & Lucas, J.W. (1980). Observations on the inter-troop movement of adult vervet monkeys (*Cercopithecus aethiops*). *Folia Primatologica*, 33, 220–235.

Henzi, S.P., Forshaw, N., Boner, R., Barrett, L. & Lusseau, D. (2013). Scalar social dynamics in female vervet monkey cohorts. *Phiosophical Transactions of the Royal Society B*, 368(1618), 20120351.

Henzi, S.P., Hetem, R., Fuller, A., Maloney, S., Young, C., Mitchell, D., Barrett, L. & McFarland, R. (2017). Consequences of sex-specific sociability for thermoregulation in male vervet monkeys during winter. *Journal of Zoology*, 302(3), 193–200.

Henzi, S.P., Lusseau, D., Weingrill, T., Van Schaik, C.P. & Barrett, L. (2009). Cyclicity in the structure of female baboon social networks. *Behavioral Ecology and Sociobiology*, 63, 1015–1021.

Hibar, D. P., Stein, J.L., Renteria, M.E. et al. (2015). Common genetic variants influence human subcortical brain structures. *Nature*, 520, 224–229.

Higham, J.P., Semple, S., MacLarnon, A., Heistermann, M. & Ross, C. (2009). Female reproductive signaling, and male mating behavior, in the olive baboon. *Hormones & Behavior*, 55, 60–67.

Higham, J.P., Brent, L.J.N., Dubuc, C., Accamando, A.K., Engelhardt, A., Gerald, M.S., Heistermann, M. & Stevens, M. (2010). Color signal information content and the eye of the beholder: a case study in the rhesus macaque. *Behavioral Ecology*, 21, 739–746.

Higley, J.D., Mehlman, P.T., Poland, R.E., Taub, D.M., Vickers, J., Suomi, S.J. & Linnoila, M. (1996). CSF testosterone and 5-HIAA correlate with different types of aggressive behaviors. *Biological Psychiatry*, 40(11), 1067–1082.

Hill, C.M. & Webber, A.D. (2010). Perceptions of nonhuman primates in human–wildlife conflict scenarios. *American Journal of Primatology*, 72, 919–924.

Hill, G.E. (2006). Female mate choice for ornamental coloration. In G.E. Hill & K.J. McGraw, eds., *Bird Coloration, Vol. 2: Function and Evolution*. Cambridge, MA: Harvard University Press, pp. 137–200.

Hill, R.A. & Dunbar, R. (1998). An evaluation of the roles of predation rate and predation risk as selective pressures on primate grouping behavior. *Behaviour*, 135, 411–430.

Hill, R.A., Barrett, L., Gaynor, D., Weingrill, T., Dixon, P., Payne, H. & Henzi, S.P. (2003). Day length, latitude and behavioural (in) flexibility in baboons (Papio cynocephalus ursinus). *Behavioral Ecology and Sociobiology*, 53, 278–286.

Hill, W.C.O. (1966). *Primate Comparative Anatomy and Taxonomy. VI. Catarrhini, Cercopithecoidea, Cercopithecinae.* Edinburgh, UK: Edinburgh University Press.

Hirsch, V.M., McGann, C., Dapolito, G. et al. (1993). Identification of a new subgroup of SIVagm in tantalus monkeys. *Virology*, 197, 426–430.

Hirsch, V.M., Olmsted, R.A., Murphey-Corb, M., Purcell, R.H. & Johnson, P.R. (1989). An African primate lentivirus (SIVsm) closely related to HIV-2. *Nature*, 339, 389–392.

Hockings, K.J., Yamakoshi, G., Kabasawa, A. & Matsuzawa, T. (2010). Attacks on local persons by chimpanzees in Bossou, Republic of Guinea, long-term perspectives. *American Journal of Primatology*, 72, 887–896.

Hoffman, M.T., Carrick, P.J., Gillson, L. & West, A.G. (2009). Drought, climate change and vegetation response in the succulent karoo, South Africa. *South African Journal of Science*, 105, 54–60.

Holmes, D.J. & Sherry, D. (1997). Selected approaches to using individual variation for understanding mammalian, life history evolution. *Journal of Mammalogy*, 78(2), 311–319.

Hope, K., Goldsmith, M.L. & Graczyk, T. (2004). Parasitic health of olive baboons in Bwindi Impenetrable National Park, Uganda. *Veterinary Parasitology*, 122(2), 165–170.

Hoppitt, W. & Laland, K.N. (2008). Social processes influencing learning in animals, a review of the evidence. *Advances in the Study of Behavior*, 38, 105–165.

Horrocks, J.A. (1982). Study of feral monkeys in Barbados. *Laboratory Primate Newsletter*, 21, 15–16.

(1984). Aspects of the behavioural ecology of Cercopithecus aethiops sabaeus in Barbados, West Indies. PhD Dissertation. Cave Hill, Barbados: University of the West Indies.

(1986). Life history characteristics of a feral population of vervets (Cercopithecus aethiops sabaeus) in Barbados. *International Journal of Primatology*, 7, 31–47.

Horrocks, J.A. & Baulu, J. (1988). Effects of trapping on the vervet (Cercopithecus aethiops sabaeus) in Barbados. *American Journal of Primatology*, 15, 223–233.

(1994). Food competition between vervets (*Cercopithecus aethiops sabaeus*) and farmers in Barbados, implications for management. *Revue d'Ecologie, La Terre et la Vie*, 49, 281–294.

Horrocks, J.A. & Hunte, W. (1983a). Maternal rank and offspring rank in vervet monkeys. *Animal Behavior*, 31, 772–782.

(1983b). Rank relations in vervet sisters. *American Naturalist*, 122, 417–421.

(1986). Sentinel behaviour in vervet monkeys, who sees whom first? *Animal Behaviour*, 34, 1566–1567.

(1993). Interactions between juveniles and adult males in vervets, implications for adult male turnover. In M.E. Pereira & L.A. Fairbanks, eds., *Juvenile Primates, Life History, Development and Behavior*. New York: Oxford University Press, pp. 228–239.

Hosseini, E., Grootaert, C., Verstraete, W. & Van de Wiele, T. (2011). Propionate as a health-promoting microbial metabolite in the human gut. *Nutrition Reviews*, 69(5), 245–258.

Howell, N. (1979). *Demography of the Dobe!Kung*. New York: Academic Press.

Hu, F., van Dam, R. & Liu, S. (2001). Diet and risk of type II diabetes: the role of types of fat and carbohydrate. *Diabetologia*, 44, 805–817.

Huang, Y.S., Ramensky, V., Service, S.K. et al. (2015) Sequencing strategies and characterization of 721 vervet monkey genomes for future genetic analyses of medically relevant traits. *BMC Biology*, 13, 41.

Huck, M., Rotundo, M. & Fernandez-Duque, E. (2011). Growth and development in wild owl monkeys (*Aotus azarai*) of Argentina. *International Journal of Primatology*, 32, 1133–1152.

Hunt, J., Bussiere, L.F., Jennions, M.D. & Brooks, R. (2004). What is genetic quality? *Trends in Ecology and Evolution*, 19(6), 329–333.

Hunte, W. & Horrocks, J.A. (1987). Kin and non-kin interventions in the aggressive disputes of vervet monkeys. *Behavioral Ecology and Sociobiology*, 20(4), 257–263.

Huxley, J.S. (1932). *Problems of Relative Growth*. London, UK: Methuen & Co.

Isbell, L.A. (1990). Sudden short-term increase in mortality of vervet monkeys (*Cercopithecus aethiops*) due to leopard predation in Amboseli National Park, Kenya. *American Journal of Primatology*, 21, 41–52.

(1991). Contest and scramble competition, patterns of female aggression and ranging behavior among primates. *Behavioral Ecology*, 2, 143–155.

(1994a). Predation on primates, ecological patterns and evolutionary consequences. *Evolutionary Anthropology*, 3, 61–71.

(1994b). The vervets' year of doom. *Natural History*, 103(8), 48–55.

(1995). Seasonal and social correlates of changes in hair, skin, and scrotal condition in vervet monkeys (*Cercopithecus aethiops*) of Amboseli National Park, Kenya. *American Journal of Primatology*, 36, 61–70.

(2004). Is there no place like home? Ecological bases of dispersal in primates and their consequences for the formation of kin groups. In B. Chapais & C. Berman, eds., *Kinship and Behavior in Primates*. New York: Oxford University Press, pp. 71–108.

(2006). Snakes as agents of evolutionary change in primate brains. *Journal of Human Evolution*, 51, 1–35.

(2009). *The Fruit, the Tree, and the Serpent: Why We See So Well*. New York: Harvard University Press.

Isbell, L.A. & Bidner, L.R. (2016). Vervet monkey (*Chlorocebus pygerythrus*) alarm calls to leopards (*Panthera pardus*) function as a predator deterrent. *Behaviour*, 153, 591–606.

Isbell, L.A., Bidner, L.R., Van Cleave, E.K., Matsumoto-Oda, A. & Crofoot, M.C. (2018). GPS-identified vulnerabilities of savannah-woodland primates to leopard predation and their implications for early hominins. *Journal of Human Evolution*, 118, 1–13.

Isbell, L.A. & Enstam, K.L. (2002). Predator (in)sensitive foraging in sympatric female vervets (*Chlorocebus aethiops*) and patas monkeys (*Erythrocebus patas*): a test of ecological models of group dispersion. In L.E. Miller, ed., *Eat or Be Eaten, Predator Sensitive Foraging in Nonhuman Primates*. New York: Cambridge University Press, pp. 154–168.

Isbell, L.A. & Jaffe, K.L.E. (2013). *Chlorocebus pygerythrus* vervet monkey. In T.M. Butynski, J.S. Kingdon & J. Kalina, eds., *The Mammals of Africa: Vol. II Primates*. London, UK: Bloomsbury Publishing, pp. 277–283.

Isbell, L.A. & Pruetz, J.D. (1998). Differences between vervets (*Cercopithecus aethiops*) and patas monkeys (*Erythrocebus patas*) in agonistic interactions between adult females. *International Journal of Primatology*, 19, 837–855.

Isbell, L.A. & Van Vuren, D. (1996). Differential costs of locational and social dispersal and their consequences for female group-living primates. *Behaviour*, 133, 1–36

Isbell, L. & Young, T. (1993a). Social and ecological influences on activity budgets of vervet monkeys, and their implications for group living. *Behavioral Ecology and Sociobiology*, 32, 377–385.

Isbell, L.A. & Young, T.P. (1993b). Human presence reduces predation in a free-ranging vervet monkey population in Kenya. *Animal Behavior*, 45, 1233–1235.

Isbell, L.A., Cheney, D.L. & Seyfarth, R.M. (1990). Costs and benefits of home range shifts among vervet monkeys (*Cercopithecus aethiops*) in Amboseli National Park, Kenya. *Behavioral Ecology and Sociobiology*, 27, 351–358.

(1991). Group fusions and minimum group sizes in vervet monkeys (*Cercopithecus aethiops*). *American Journal of Primatology*, 25, 57–65.

(1993). Are immigrant vervet monkeys, *Cercopithecus aethiops*, at greater risk of mortality than residents? *Animal Behavior*, 45, 729–734.

(2002). Why vervet monkeys (*Cercopithecus aethiops*) live in multimale groups. In M. Glenn & M. Cords, eds., *The Guenons, Diversity and Adaptation in African Monkeys*. New York: Kluwer Academic/Plenum Publishers, pp. 173–187.

Isbell, L.A, Pruetz, J.D. & Young, T.P. (1998). Movements of vervets (*Cercopithecus aethiops*) and patas monkeys (*Erythrocebus patas*) as estimators of food resource size, density, and distribution. *Behavioral Ecology and Sociobiology*, 42, 123–133.

Isbell, L.A., Young, T.P., Jaffe, K.E., Carlson, A.A. & Chancellor, R.L. (2009). Demography and life histories of sympatric patas monkeys (*Erythrocebus patas*) and vervets (*Cercopithecus aethiops*) in Laikipia, Kenya. *International Journal of Primatology*, 30, 103–124.

Ishimoto, G., Toyomazu, T. & Uemura, K. (1967). Serum transferrins of Japanese macaques: comparison with other species of monkeys. *Primates*, 8, 29–34.

Jablonski, N.G. (2002). Fossil Old World monkeys: the late Neogene radiation. In W.C. Hartwig, ed., *The Primate Fossil Record*. New York: Cambridge University Press, pp. 255–299.

Jablonski, N.G. & Frost, S. (2010). Cercopithecoidea. In L. Werderlin & W.J. Sanders, eds., *Cenozoic Mammals of Africa*. Los Angeles, CA: University of California Press, pp. 393–428.

Jackson, G. & Gartlan, J.S. (1965). The flora and fauna of Lolui Island, Lake Victoria: a study of vegetation, men and monkeys. *Journal of Ecology*, 53(3), 573–598.

Jacobus, S. & Loy, J. (1981). The grimace and gecker, a submissive display among patas monkeys. *Primates*, 22, 393–398.

Jacquelin, B., Mayau, V., Targat, B. et al. (2009). Nonpathogenic SIV infection of African green monkeys induces a strong but rapidly controlled type I IFN response. *Journal of Clinical Investigation*, 119, 3544–3555.

Jaffe, K.E. & Isbell, L.A. (2009). After the fire: benefits of reduced ground cover for vervet monkeys (*Cercopithecus aethiops*). *American Journal of Primatology*, 71, 252–260.

(2010). Ranging and social behavior after a predator-induced group fusion in wild vervet monkeys (*Cercopithecus aethiops*) in Laikipia, Kenya. *American Journal of Primatology*, 72, 634–644.

Janson, C. (2000). Spatial movement strategies: theory, evidence, and challenges. In S. Boinski & P.A. Garber, eds., *On the Move: How and Why Animals Travel in Groups*. Chicago: University of Chicago Press, pp. 165–203.

Janson, C.H. & van Schaik, C.P. (1988). Recognizing the many faces of primate food competition, methods. *Behaviour*, 105, 165–186.

(1993). Ecological risk aversion in juvenile primates: slow and steady wins the race. In M.E. Pereira & L.A. Fairbanks, eds., *Juvenile Primates*. New York: Oxford University Press, pp. 57–74.

Jarman, P. (1974). The social organisation of antelope in relation to their ecology. *Behaviour*, 48(1), 215–267.

Jasinska, A.J., Lin, M.K., Service, S. et al. (2012) A nonhuman primate system for large-scale genetic studies of complex traits. *Human Molecular Genetics*, 21(15), 3307–3316.

Jasinska, A.J., Schmitt, C.A., Service, S.K. et al. (2013). Systems biology of the vervet monkey. *ILAR Journal*, 54, 122–143.

Jasinska, A.J., Service, S., Levinson, M. et al. (2007). A genetic linkage map of the vervet monkey (*Chlorocebus aethiops sabaeus*). *Mammalian Genome*, 18, 347–360.

Jasinska, A.J., Service, S., Choi, O.W. et al. (2009). Identification of brain transcriptional variation reproduced in peripheral blood, an approach for mapping brain expression traits. *Human Molecular Genetics*, 18, 4415–4427.

Jasinska, A.J., Zelaya, I., Service, S.K. et al. (2017). Genetic variation and gene expression across multiple tissues and developmental stages in a nonhuman primate. *Nature Genetics*, 49(12), 1714–1721.

Jin, M.J., Hui, H., Robertson, D.L, Muller, M.C., Barre-Sinoussi, F., Hirsch, V.M., Allan, J.S., Shaw, G.M., Sharp, P.M. & Hahn, B.H. (1994a). Mosaic genome structure of simian immunodeficiency virus from West African green monkeys. *EMBO Journal*, 13, 2935–2947.

Jin, M.J., Rogers, J., Phillips-Conroy, J.E., Allan, J.S., Desrosiers, R.C., Shaw, G.M., Sharp, P.M. & Hahn, B.H. (1994b). Infection of a yellow baboon with simian immunodeficiency virus from African green monkeys, evidence for cross-species transmission in the wild. *Journal of Virology*, 68, 8454–8460.

Johnson, S.E. (2003). Life history and the competitive environment: trajectories of growth, maturation and reproductive output among chacma baboons. *American Journal of Physical Anthropology*, 120(1), 83–98.

Jolly, C.J. (1993). Species, subspecies and baboon systematics. In W.H. Kimbel & L.B. Martin, eds., *Species, Species Concepts and Primate Evolution*. New York: Plenum Press, pp. 67–108.

Jolly, C.J. & Brett, F.L. (1973). Genetic markers and baboon biology. *Journal of Medical Primatology*, 2, 85–99.

Jolly, C.J. & Phillips-Conroy, J.E. (2003). Testicular size, mating system, and maturation schedules in wild Anubis and hamadryas baboons. *International Journal of Primatology*, 24, 125–142.

Jolly, C.J., Phillips-Conroy, J.E., Turner, T.R., Broussard, S. & Allan, J.S. (1996). SIV-agm incidence over two decades in a natural population of Ethiopian grivet monkeys (*Cercopithecus aethiops aethiops*). *Journal of Medical Primatology*, 25(2), 78–83.

Jolly, C.J., Turner, T.R., Socha, W.W. & Weiner, A.S. (1977) Human-type A-B-O blood group antigens of Ethiopian vervet monkeys (*Cercopithecus aethiops*) in the wild. *Journal of Medical Primatology*, 6, 54–57.

Jones, G.E. & Scott, G.R. (1970). Vervet monkey disease, a new zoonosis. *Tropical Animal Health and Production*, 2(1), 35–43.

Jones-Engel, L., Engel, G.A., Schillaci, M.A., Babo, R. & Froehlich, J. (2001). Detection of antibodies to selected human pathogens among wild and pet macaques (*Macaca tonkeana*) in Sulawesi, Indonesia. *American Journal of Primatology*, 54(3), 171–178.

Jones-Engel, L., Schillaci, M.A. & Engel, G.A. (2011). Human–nonhuman primate interactions, an ethnoprimatological approach. In J.M. Setchell & D.J. Curtis, eds., *Field and Laboratory Methods in Primatology, A Practical Guide*, 2nd edn. Cambridge: Cambridge University Press, pp. 15–24.

Josephs, N., Bonnell, T., Dostie, M., Barrett, L. & Henzi, S.P. (2016). Working the crowd: sociable vervets benefit by reducing exposure to risk. *Behavioral Ecology*, 27, 988–994.

Jost Robinson, C.A. & Remis, M.J. (2014). Entangled realms, hunters and hunted in the Dzanga Sangha Dense Forest Reserve (RDS), Central African Republic. *Anthropology Quarterly*, 87, 613–636.

Kalin, N.H. (2004). Studying non-human primates, a gateway to understanding anxiety disorders. *Psychopharmacology Bulletin*, 38, 8–13.

Kanthaswamy, S. & Smith, D.G. (2002). Population subdivision and gene flow among wild orangutans. *Primates*, 43(4), 315–327.

Kaplan, J.R. & Zucker, E.L. (1980). Social organization in a group of free-ranging patas monkeys. *Folia Primatologica*, 34, 196–213.

Kappeler, P.M., Pereira, M.E. & van Schaik, C.P. (2003). Primate life histories and socioecology. In P.M. Kappeler & M.E. Pereira, eds., *Primate Life Histories and Socioecology*. Chicago: Chicago University Press, pp. 1–23.

Kapusinszky, B., Mulvaney, U., Jasinska, A.J. et al. (2015). Local virus extinctions following a host population bottleneck. *Journal of Virology*, 89, 8152–8161.

Kavanagh, M. (1978). The diet and feeding behaviour of *Cercopithecus aethiops tantalus*. *Folia Primatologica*, 30, 30–63.

Kavanagh, K., Fairbanks, L.A., Bailey, J.N. et al. (2007). Characterization and heritability of obesity and associated risk factors in vervet monkeys. *Obesity*, 15, 1666–1674.

Kawai, M. (1965). Newly-acquired pre-cultural behavior of the natural troop of Japanese monkeys on Koshima Islet. *Primates*, 6, 1–30.

Kawamoto, Y., Ischak, T.M. & Supriatna, J. (1984). Genetic variations within and between troops of the crab-eating macaque (*Macaca fascicularis*) on Sumatra, Java, Bali, Lombok and Sumbawa, Indonesia. *Primates*, 25(2), 131–159.

Kawamoto, Y., Shotake, T. & Nozawa, K. (1982). Genetic differentiation among three genera of family Cercopithecidae. *Primates*, 23(2), 272–286.

Keane, B., Dittus, W.P.J. & Melnick, D.J. (1997). Paternity assessment in wild groups of toque macaques *Macaca sinica* at Polonnaruwa, Sri Lanka using molecular markers. *Molecular Ecology*, 6(3), 267–282.

Keddy, A.C. (1986). Female mate choice in vervet monkeys *(Cercopithecus aethiops sabaeus)*. *American Journal of Primatology*, 10, 125–134.

Keddy Hector, A.C. & Raleigh, M.J. (1992). The effects of temporary removal of the alpha male on the behavior of subordinate male vervet monkeys. *American Journal of Primatology*, 26, 77–87.

Keele, B.F., Giorgi, E.E., Salazar-Gonzalez, J.F. et al. (2008). Identification and characterization of transmitted and early founder virus envelopes in primary HIV-1 infection. *Proceedings of the National Academy of Science of the United States of America*, 105, 7552–7557.

Keele, B.F., Li, H., Learn, G.H. et al. (2009). Low dose rectal inoculation of rhesus macaques by SIVsmE660 or SIVmac251 recapitulates human mucosal infection by HIV-1. *Journal of Experimental Medicine*, 206, 1117–1134.

Kelaita, M.A. (2015). Applications of genomic methods to studies of wild primate populations. In R. Duggirala, L. Almasy, S. Williams-Blangero, F.D. Paul & C. Kole, eds., *Genome Mapping and Genomics in Human and Non-Human Primates*. Berlin, Germany: Springer, pp. 103–112.

Kessel, A. & Brent, L. (2001). The rehabilitation of captive baboons. *Journal of Medical Primatology*, 30(2), 71–80.

Kessel, A.L. & Brent, L. (1998). Cage toys reduce abnormal behavior in individually housed pigtail macaques. *Journal of Applied Animal Welfare Science*, 1(3), 227–234.

Kleiber, M. (1961). *The Fire of Life: An Introduction to Animal Energetics*. New York: John Wiley & Sons, Inc.

Kilmer, J.T. & Rodríguez, R.L. (2017). Ordinary least squares regression is indicated for studies of allometry. *Journal of Evolutionary Biology*, 30(1), 4–12.

King, S.J., Morelli, T.L., Arrigo-Nelson, S., Ratelolahy, F.J., Godfrey, L.R., Wyatt, J., Tecot, S., Jernvall, J. & Wright, P.C. (2011). Morphometrics and pattern of growth in wild sifakas (*Propithecus edwardsi*) at Ranomafana National Park, Madagascar. *American Journal of Primatology*, 73(2), 155–172.

Kingdon J. (1971). *East African Mammals*, Vol. 1. New York: Academic Press.

Kingdon, J. (1988). What are face patterns and do they contribute to reproductive isolation in guenons? In A. Gautier-Hion, F. Bourliere, J.P. Gautier-Hion & J. Kingdon, eds., *A Primate Radiation: Evolutionary Biology of the African Guenons*. Cambridge: Cambridge University Press, pp. 227–245.

(1992). Facial patterns as signals and masks. In S. Jones, R. Martin & D. Pilbeam, eds., *The Cambridge Encyclopedia of Human Evolution*. Cambridge: Cambridge University Press, pp. 161–165.

Kingdon, J. & Largen, M. J. (1997). The kingdom field guide to African mammals. *Zoological Journal of the Linnean Society*, 120(4), 479.

Kisidayova, S., Varadyova, Z., Pristas, P. et al. (2009). Effects of high- and low-fiber diets on fecal fermentation and fecal microbial populations of captive chimpanzees. *American Journal of Primatology*, 71, 548–557.

Kluge, M. & Ting, I. (1978). *Crassulacean Acid Metabolism: An Ecological Analysis*. Berlin, Germany: Springer-Verlag.

Knight, J. (1999). Monkeys on the move, the natural symbolism of people–macaque conflict in Japan. *The Journal of Asian Studies*, 58, 622–647.

(2000). Introduction. In J. Knight, ed., *Natural Enemies: People–Wildlife Conflicts in Anthropological Perspective*. London: Routledge, pp. 1–35.

Kodric-Brown, A., Sibly, R.M. & Brown, J. H. (2006). The allometry of ornaments and weapons. *Proceedings of the National Academy of Sciences of the United States of America*, 103(23), 8733–8738.

Korstjens, A.H., Lehmann, J. & Dunbar, R.I.M. (2010). Resting time as an ecological constraint on primate biogeography. *Animal Behaviour*, 79, 361–374.

Kowalewski, M.M., Salzer, J.S., Deutsch, J.C., Rano, M., Kuhlenschmidt, M.S. & Gillespie, T.R. (2011). Black and gold howler monkeys (*Alouatta caraya*) as sentinels of ecosystem health: patterns of zoonotic protozoa infection relative to degree of human–primate conflict. *American Journal of Primatology*, 73(1), 75–83.

Krige, P.D. & Lucas, J.W. (1974). Aunting behaviour in an urban troop of *Cercopithecus aethiops*. *Journal of Behavioural Science*, 2, 55–61.

Kuntz, R.E. & Myers, B.J. (1966). Parasites of baboons (*Papio doguera* (Pucheran, 1856)) captured in Kenya and Tanzania, East Africa. *Primates*, 7, 27–32.

Kuokkanen, S., Polotsky, A.J., Chosich, J., Bradford, A.P., Jasinska, A., Phang, T., Santoro, N. & Appt, S.E. (2016). Corpus luteum as a novel target of weight changes that contribute to impaired female reproductive physiology and function. *Systems Biology in Reproductive Medicine*, 62(4), 227–242.

Kurita, H., Suzumura, T., Kanchi, F. & Hamada, Y. (2012). A photogrammetric method to evaluate nutritional status without capture in habituated free-ranging Japanese macaques (*Macaca fuscata*), a pilot study. *Primates*, 53, 7–11.

Kuzawa, C.W. (1998). Adipose tissue in human infancy and childhood: an evolutionary perspective. *Yearbook of Physical Anthropology*, 41, 177–209.

Lack, D. (1947). The significance of clutch size. *Ibis*, 89, 302–352.

Laland, K.N. (2004). Social learning strategies. *Learning & Behavior*, 32, 4–14.

Laland, K.N. & Janik, V. (2006). The animal cultures debate. *Trends in Ecology & Evolution*, 21, 542–547.

Lancaster, J.B. (1971). Play-mothering: the relations between juvenile females and young infants among free-ranging vervet monkeys (*Cercopithecus aethiops*). *Folia Primatologica*, 15, 161–182.

Lane, K.E., Lute, M., Rompis, A., Wandia, I.N., Putra, I.A., Hollocher, H. & Fuentes, A. (2010). Pests, pestilence, and people: the long-tailed macaque and its role in the cultural complexities of Bali. In S. Gursky-Doyen & J. Supriatna, eds., *Indonesian Primates*. New York, NY: Springer, pp. 235–248.

Langergraber, K.E, Siedel, H., Mitani, J.C., Wrangham, R.W., Reynolds, V., Hunt, K. & Vigilant, L. (2007). The genetic signature of sex-biased migration in patrilocal chimpanzees and humans. *PLoS One*, 2(10), 1–7.

Larsen, N., Vogensen, F.K., van den Berg, F.W.J. et al. (2010). Gut microbiota in human adults with type 2 diabetes differs from non-diabetic adults. *PLoS One*, 5(2), e9085.

Laudenslager, M.L., Jorgensen, M.J., Grzywa, R. & Fairbanks, L.A. (2011). A novelty seeking phenotype is related to chronic hypothalamic–pituitary–adrenal activity reflected by hair cortisol. *Physiology & Behavior*, 104, 291–295.

Launhardt, K , Epplen, C., Epplen, J.T. & Winkler, P. (1998). Amplification of microsatellites adapted from human systems in faecal DNA of wild Hanuman langurs (*Presbytis entellus*). *Electrophoresis*, 19(8–9), 1356–1361.

Laviola, G., Macrì, S., Morley-Fletcher, S. & Adriani, W. (2003). Risk-taking behavior in adolescent mice: psychobiological determinants and early epigenetic influence. *Neuroscience and Biobehavioral Reviews*, 27, 19–31.

Lawler, R.R., Richard, A.F. & Riley, M.A. (2003). Genetic population structure of the white sifaka (*Propithecus verrauxi verrauxi*) at Beza Mahafaly Special Reserve, southwest Madagascar (1992–2001). *Molecular Ecology*, 12(9), 2307–2317.

Le, Q.V., Isbell, L.A., Nguyen, M.N., Matsumoto, J., Hori, E., Maior, R.S., Tomaz, C., Tran, A.H., Ono, T. & Nishijo, H. (2013). Pulvinar neurons reveal neurobiological evidence of past selection for rapid detection of snakes. *Proceedings of the National Academy of Sciences of the United States of America*, 110, 19000–19005.

Le, Q.V., Isbell, L.A., Matsumoto, J., Quang, L.V., Hori, E., Tran, A.H., Maior, R.S., Tomaz, C., Ono, T. & Nishijo, H. (2014). Monkey pulvinar neurons fire differentially to snake postures. *PLoS One*, 9, e114258.

Leakey, M. (1988). Fossil evidence for the evolution of the guenons. In A. Gautier-Hion, F. Bourliere, J.P. Gautier-Hion & J. Kingdon, eds., *A Primate Radiation: Evolutionary Biology of the African Guenons*. Cambridge: Cambridge University Press, pp. 7–12.

Leakey, M.G., Teaford, M.F. & Ward, C.V. (2003). Cercopithecidae from lothagam. In M. Leakey, ed., *Lothagam: The Dawn of Humanity in Eastern Africa*. New York: Columbia University Press, New York, pp. 201–248.

Ledbetter, D.H. (1981). Chromosomal evolution and speciation of the genus Cercopithecus (Primates, Cercopithecinae). PhD thesis. Austin, TX: University of Texas.

Lee, P.C. (1984). Early infant development and maternal care in free-ranging vervet monkeys. *Primates*, 25(1), 36–47.

(2010). Sharing space, can ethnoprimatology contribute to the survival of non-human primates in human-dominated globalized landscapes? *American Journal of Primatology*, 72, 925–931.

Lee, P.C. & Kappeler, P.M. (2003). Socioecological correlates of phenotypic plasticity of primate life histories. In P.M. Kappeler & M. E. Pereira, eds., *Primate Life Histories and Socioecology*. Chicago: Chicago University Press, pp. 41–65.

Lee, P.C. & Priston, N.E.C. (2005). Human attitudes to primates: perception of pests, conflict and consequences for conservation. In J.D. Paterson & J. Wallis, eds., *Commensalism and Conflict: The Human–Primate Interface*. Norman, OK: American Society of Primatologists, pp. 1–23.

Lee-Thorp, J.A. & van der Merwe, N.J. (1987). Carbon isotope analysis of fossil bone apatite. *South African Journal of Science*, 83, 712–715.

Legesse, M. & Erko, B. (2004). Zoonotic intestinal parasites in *Papio anubis* (baboon) and *Cercopithecus aethiops* (vervet) from four localities in Ethiopia. *Acta Tropica*, 90(3), 231–236.

Leigh, S.R. (1992). Patterns of variation in the ontogeny of primate body size dimorphism. *Journal of Human Evolution*, 23(1), 27–50.

Leonard, W.R. (1989). Nutritional determinants of high-altitude growth in Nunoa, Peru. *American Journal of Physical Anthropology*, 80(3), 341–352.

Leonard, W.R., Robertson, M.L., Snodgrass, J.J. & Kuzawa, C.W. (2003). Metabolic correlates of hominid brain evolution. *Comparative Biochemistry and Physiology A*, 135, 5–15.

Leonard, W.R., Snodgrass, J.J. & Robertson, M.L. (2007). Effects of brain evolution on human nutrition and metabolism. *Annual Review of Nutrition*, 27, 311–327.

Lernould, J.M. (1988). Classification and geographical distribution of guenons: a review. In A. Gautier-Hion, F. Bourliere, J.P. Gautier-Hion & J. Kingdon, eds., *A Primate Radiation: Evolutionary Biology of the African Guenons.* Cambridge: Cambridge University Press, pp. 54–78.

Lewis, L.S., Rogers, J.A. & Turner, T.R. (1981). Some female reproductive statistics for wild-caught vervet monkeys (Cercopithecus aethiops pygerythrus). *American Journal of Physical Anthropology,* 54, 246.

Ley, R.E. (2010). Obesity and the human microbiome. *Current Opinion in Gastroenterology,* 26(1), 5–11.

Li, D., Ren, B., Grueter, C.C., Li, B. & Li, M. (2010). Nocturnal sleeping habits of the Yunnan snub-nosed monkey in Xiangguqing, China. *American Journal of Primatology,* 72, 1092–1099.

Li, H.P., Meng, S.J., Men, Z.M., Fu, Y.X. & Zhang. Y.P. (2003). Genetic diversity and population history of golden monkeys (*Rhinopithecus roxellana*). *Genetics,* 164(1), 269–275.

Lilly, A.A., Mehlman, P.T. & Doran, D. (2002). Intestinal parasites in gorillas, chimpanzees, and humans at Mondika Research site, Dzanga-Ndoki National Park, Central African Republic. *International Journal of Primatology,* 23(3), 555–573.

Ling, B., Apetrei, C., Pandrea, I., Veazey, R.S., Lackner, A.A., Gormus, B. & Marx, P.A. (2004). Classic AIDS in a sooty mangabey after an 18-year natural infection. *Journal of Virology,* 78, 8902–8908.

Locke, A.E., Kahali, B. & Berndt, S.I. (2015). Genetic studies of body mass index yield new insights for obesity biology. *Nature,* 518, 197–206.

Lorenz, J., Grobler, J.P., MacAuliffe Dore, K., Freimer, N., Jasinska, A. & Turner, T.R. (2010). Genetic variation among geographically widespread populations of vervets (*Chlorocebus aethiops*) in southern and eastern Africa. *American Journal of Physical Anthropology,* 141, 157–158.

Loudon, J.E., Grobler. J.P., Sponheimer. M., Moyer. K., Lorenz. J.G. & Turner, T.R. (2014). Using stable carbon and nitrogen isotope compositions of vervet monkey (*Chlorocebus pygerythrus*) to examine questions in ethnoprimatology. *PLoS One,* 9, e100758.

Loudon, J.E., Howells, M.E. & Fuentes, A. (2006). The importance of integrative anthropology: a preliminary investigation employing primatological and cultural anthropological data collection methods in assessing human–monkey coexistence in Bali, Indonesia. *Ecological and Environmental Anthropology,* 2, 2–13.

Loudon, J.E., Sandberg, P.A., Wrangham, R.W., Fahey, B. & Sponheimer, M. (2016). The stable isotope ecology of *Pan* in Uganda and beyond. *American Journal of Primatology,* 78(10), 1070–1085.

Loudon, J.E., Sponheimer, M., Sauther, M.L. & Cuozzo, F.P. (2007). Intraspecific variation in hair δ^{13}C and δ^{15}N values of ring-tailed lemurs (*Lemur catta*) with known individual histories, behavior, and feeding ecology. *American Journal of Physical Anthropology*, 133, 978–985.

Loy, J. & Harnois, M. (1988). An assessment of dominance and kinship among patas monkeys. *Primates*, 29, 331–342.

Lozupone, C., Stombaugh, J., Gonzalez, A. et al. (2012). The vaginal microbiome, rethinking health and diseases. *Annual Review of Microbiology*, 66, 371–389.

Lu, A., Bergman, T.J., McCann, C., Stinespring-Harris, A. & Beehner, J.C. (2016). Growth trajectories in wild geladas (*Theropithecus gelada*). *American Journal of Primatology*, 78, 707–719.

Lubbe, A., Hetem, R.S., McFarland, R., Barrett, L., Henzi, S.P., Mitchell, D., Meyer, L.C.R., Maloney, S.K. & Fuller, A. (2014). Thermoregulatory plasticity in free-ranging vervet monkeys, Chlorocebus pygerythrus. *Journal of Comparative Physiology B*, 184, 799–809.

Lucotte, G., Gauteau, C., Galat, G. & Galat-Luong, A. (1982). Electrophoretic polymorphism in different subspecies of *Cercopithecus aethiops*. *Folia Primatologica*, 38(3–4), 183–195.

Lukas, D., Reynolds, V., Boesch, C. & Vigilant, L. (2005). To what extent does living in a group mean living with kin? *Molecular Ecology*, 14(7), 2181–2196.

Ma, B., Forney, L.J. & Ravel, J. (2012). Vaginal microbiome: rethinking health and disease. *Annual Review of Microbiology*, 66, 371–389.

Ma, D., Jasinska, A., Kristoff, J. et al. (2014). Factors associated with SIV transmission in a natural African nonhuman primate host in the wild. *Journal of Virology*, 88, 5687–5705.

Ma, J., Prince, A.L., Bader, D. et al. (2014). High-fat maternal diet during pregnancy persistently alters the offspring microbiome in a primate model. *Nature Communications*, 5, 3889.

Ma, D. Jasinska, A.J., Kristoff, J. et al. (2013). SIVagm infection in wild African green monkeys from South Africa: epidemiology, natural history, and evolutionary considerations. *PLoS Pathogens*, 9, e1003011.

MacIntosh, A.J.J., Jacobs, A., Garcia, C. et al. (2012). Monkeys in the middle, parasite transmission through the social network of a wild primate. *PLoS One*, 7, e51144.

Maestripieri, D., Hoffman, C.L., Anderson, G.M., Carter, C.S. & Higley, J.D. (2009). Mother–infant interactions in free-ranging rhesus macaques, relationships between physiological and behavioral variables. *Physiology & Behavior*, 96, 613–619.

Majolo, B., McFarland, R., Young, C. & Qarro, M. (2013). The effect of climatic factors on the activity budgets of a temperate primate, the Barbary macaque (Macaca sylvanus). *International Journal of Primatology*, 34, 500–514.

Marketon, J.I.W. & Glaser, R. (2008). Stress hormones and immune function. *Cellular Immunology*, 252, 16–26.

Marmoset Genome Sequencing and Analysis Consortium (2014). The common marmoset genome provides insight into primate biology and evolution. *Nature Genetics*, 46, 850–857.

Martin, R.D. & MacLarnon, A.M. (1988). Quantitative comparisons of the skull and teeth in guenons. In A. Gautier-Hion, F. Bourliere, J.P. Gautier-Hion & J. Kingdon, eds., *A Primate Radiation: Evolutionary Biology of the African Guenons*. Cambridge: Cambridge University Press, pp. 160–183.

Martin, R.D., Dixon, A.F. & Wickings, E.J. (1992). *Paternity in Primates: Genetic Tests and Theories: Implications of Human DNA Fingerprinting*. Basel, Switzerland: Karger.

Martinez-Mota, R., Valdespino, C., Sanchez-Ramos, M.A. & Serio-Silva, J.C. (2007). Effects of forest fragmentation on the physiological stress response of black howler monkeys. *Animal Conservation*, 10(3), 374–379.

Maslin, M.A., Shultz, S. & Trauth, M.H. (2015). A synthesis of the theories and concepts of early human evolution. *Philosophical Transactions of the Royal Society of London B: Biological Sciences*, 370, 20140064.

Mathy, J. & Isbell, L.A. (2001). The relative importance of size of food and interfood distance in eliciting aggression in captive rhesus macaques (*Macaca mulatta*). *Folia Primatologica*, 72, 268–277.

Mayr, E. (1963). *Animal Species and Evolution*. Cambridge, MA: Harvard University Press.

McCombs, M.L. & Bowman, B.H. (1970). Electrophoretic comparison of ceruloplasmin types in ten primate species. *Texas Reports on Biology and Medicine*, 28(1), 69–74.

McDermid, E.M. & Ananthakrishnan, R. (1972). Red cell enzymes and serum proteins of *Cercopithecus aethiops* (South African green monkey). *Folia Primatologica*, 17(1), 122–131.

McDermid, E.M., Vos, G.H. & Downing, H.J. (1973). Blood groups, red cell enzymes and serum proteins of baboons and vervets. *Folia Primatologica*, 19, 312–326.

McDougall, P., Forshaw, N., Barrett, L. & Henzi, S.P. (2010). Leaving home: responses to water depletion by vervet monkeys. *Journal of Arid Environments*, 74, 924–927.

McFarland, R. & Majolo, B. (2013). Coping with the cold: predictors of survival in wild Barbary macaques, Macaca sylvanus. *Biology Letters*, 9, 20130428.

McFarland, R., Barrett, L., Boner, R., Freeman, N. J. & Henzi, S. P. (2014). Behavioral flexibility of vervet monkeys in response to climatic and social variability. *American Journal of Physical Anthropology*, 154, 357–364.

McFarland, R., Fuller, A., Hetem, R.S., Mitchell, D., Maloney, S.K., Henzi, S.P. & Barrett, L. (2015). Social integration confers thermal benefits in a gregarious primate. *Journal of Animal Ecology*, 84, 871–878.

McFarland, R., Henzi, S.P., Barrett, L., Wanigaratne, A., Coetzee, E., Fuller, A., Hetem, R.S., Mitchell, D. & Maloney, S.K. (2016). Thermal consequences of increased pelt loft infer an additional utilitarian function for grooming. *American Journal of Primatology*, 78, 456–461.

McFarland, R., Hetem, R.S., Fuller, A., Mitchell, D., Henzi, S.P. & Barrett, L. (2013). Assessing the reliability of biologger techniques to measure activity in a free-ranging primate. *Animal Behaviour*, 85, 861–866.

McFarland, R., Murphy, D., Lusseau, D., Henzi, S.P., Parker, J.L., Pollet, T.V. & Barrett, L. (2017). The "strength of weak ties" among female baboons: fitness-related benefits of social bonds. *Animal Behaviour*, 126, 101–106.

McGrew, W.C., Tutin, C.E.G., Collins, D.A. & File, S.K. (1989). Intestinal parasites of sympatric *Pan troglodytes* and *Papio* spp. at two sites: Gombe (Tanzania) and Mt. Assirik (Senegal). *American Journal of Primatology*, 17(2), 147–155.

McGuire, M.T. (1974). *The St. Kitts Vervet*. New York: Karger.

McGuire, M.T. & Raleigh, M.J. (1985). Serotonin–behavior interactions in vervet monkeys. *Psychopharmacology Bulletin*, 21(3), 458–463.

McKee, J.K., von Mayer, A. & Kuykendall, K.L. (2011). New species of Cercopithecoides from Haasgat, North West Province, South Africa. *Journal of Human Evolution*, 60(1), 83–93.

McKinney, T. & Dore, K.M. (2018). The state of ethnoprimatology: its use and potential in today's primate research. *International Journal of Primatology*, 1–19.

McNamara, J.M. & Houston, A.I. (1996). State-dependent life histories. *Nature*, 380, 215–221.

McPhee, M.E. (2003). Generations in captivity increases behavioral variance, considerations for captive breeding and reintroduction programs. *Biological Conservation*, 115, 71–77.

McPherson, M., Smith-Lovin, L. & Brashears, M.E. (2006). Social isolation in America: changes in core discussion networks over two decades. *American Sociological Review*, 71, 353–375.

Medina, E. & Minchin, P. (1980). Stratification of $\delta^{13}C$ values of leaves in Amazonian rain forests. *Oecologia*, 45, 377–378.

Mellors, J.W., Rinaldo Jr., C.R., Gupta, P., White, R.M., Todd, J.A. & Kingsley, L.A. (1996). Prognosis in HIV-1 infection predicted by the quantity of virus in plasma. *Science*, 272, 1167–1170.

Melnick, D.J. & Kidd. K.K. (1983). The genetic consequences of social group fission in a wild population of rhesus monkeys (*Macaca mulatta*). *Behavioral Ecology and Sociobiology*, 12(3), 229–236.

Melnick, D.J. & Pearl, M.C. (1987). Cercopithecines in multimale groups: genetic diversity and population structure. In B.B. Smuts, D.L. Cheney, R.M. Seyfarth, R.W. Wrangham & T.T. Struhsaker, eds., *Primate Societies.* Chicago: University of Chicago Press, pp. 121–134.

Melnick, D.J., Jolly, C.J. & Kidd, K.K. (1984). The genetics of a wild population of rhesus monkeys (*Macaca mulatta*). I. Genetic variability within and between social groups. *American Journal of Physical Anthropology*, 63(4), 341–360.

Menard, N., von Segesser, F., Scheffrahn, W.,Pastorini, J., Vallet, D., Gaci, B., Martin, R.D. & Gautier-Hion, A. (2001). Is male-infant caretaking related to paternity and/or mating activities in wild Barbary macaques (*Macaca sylvanus*)? *Comptes Rendus de l'Academic des Sciences*, 324(7), 601–610.

Mendoza, S.P., Capitanio, J.P. & Mason, W.A. (2000). Chronic social stress: studies in non-human primates. In G.P. Moberg & J.A. Mench, eds., *Biology of Animal Stress: The Basic Principles and Implications for Animal Welfare.* New York: CABI Publishing, pp. 227–247.

Metcalfe, N.B. & Monaghan, P. (2001). Compensation for a bad start: grow now, pay later? *Trends in Ecology & Evolution*, 16, 254–260.

Meyer, J.S. & Bowman, R.E. (1972). Rearing experience, stress and adrenocorticosteroids in the rhesus monkey. *Physiology & Behavior*, 8(2), 339–343.

Michener, C.D. & Sokal, R.R. (1957). A quantitative approach to a problem of classification. *Evolution*, 11, 490–499.

Milton, K. & May, M.L. (1976). Body weight, diet and home range area in primates. *Nature*, 259, 459–462.

(1988). Foraging behavior and the evolution of primate cognition. In A. Whiten & R. Byrne, eds., *Machiavellian Intelligence, Social Expertise and the Evolution of Intellect in Monkeys, Apes and Humans.* Oxford, UK: Oxford University Press, pp. 285–305.

Mitani, J.C., Merriwether, D.A. & Zhang, C.B. (2000). Male affiliation, cooperation and kinship in wild chimpanzees. *Animal Behaviour*, 59(4), 885–893.

Mitchell, G. & Tokunaga, D.H. (1976). Sex differences in nonhuman primate grooming. *Behavioural Processes*, 1, 335–345.

Miththapala, S., Seidensticker, J. & O'Brien, S.J. (1996). Phylogeographic subspecies recognition in leopards (*Panthera pardus*): molecular genetic variation. *Conservation Biology*, 10(4), 1115–1132.

Mittermeier, R.A., Schwitzer, C., Rylands, A.B., Taylor, L., Chiozza, F., Williamson, E.A. & Wallis, J. (2012). *Primates in Peril, The World's Top 25 Most Endangered Primates 2012–2014.* Bristol, UK: Conservation International.

Mittermeier, R.A., Wilson, D.E. & Rylands, A.B. (2013). *Handbook of the Mammals of the World: Primates.* Barcelona, Spain: Lynx Edicions.

Montgomerie, R. (2006). Analyzing colors. In G.E. Hill & K.J. McGraw, eds., *Bird Coloration Vol. 1: Mechanisms and Measurements.* Cambridge, MA: Harvard University Press, pp. 90–147.

Moore, J.P., Kitchen, S.G., Pugach, P. & Zack, J.A. (2004). The CCR5 and CXCR4 coreceptors – central to understanding the transmission and pathogenesis of human immunodeficiency virus type 1 infection. *AIDS Research and Human Retroviruses,* 20, 111–126.

Moulin, S., Gerbault-Seureau, M., Dutrillaux, B. & Richard, F.A. (2008). Phylogenomics of African guenons. *Chromosome Research,* 16(5), 783–799.

Muriuki, S.M.K., Murugu, R.K., Munene, E., Karere, G.M. & Chai, D.C. (1998). Some gastro-intestinal parasites of zoonotic (public health) importance commonly observed in old wold non-human primates in Kenya. *Acta Tropica,* 71(1), 73–82.

Mutani, A., Rhynd, K. & Brown, G. (2003). A preliminary investigation on the gastrointestinal helminthes of the Barbados green monkey, *Cercopithecus aethiops sabaeus. Revista do Instituto de Medicina Tropical de Sao Paolo,* 45(4), 193–195.

Mzilikazi, N., Masters, J.C. & Lovegrove, B.G. (2006). Lack of torpor in free-ranging southern lesser galagos, Galago moholi: ecological and physiological considerations. *Folia Primatologica,* 77, 465–476.

Nakagawa, N. (1992). Distribution of affiliative behaviors among adult females within a group of wild patas monkeys in a nonmating, nonbirth season. *International Journal of Primatology,* 13, 73–96.

Nakagawa, S. & Schielzeth, H. (2010). Repeatability for Gaussian and non-Gaussian data: a practical guide for biologists. *Biological Reviews,* 85(4), 935–956.

Napier, P.H. (1981). *Catalogue of Primates in the British Museum (Natural History) and Elsewhere in the British Isles, Part II: Family Cercopithecidae, Subfamily Cercopithecinae.* London, UK: British Museum (Natural History).

Nee, S., Colegrave, N., West, S.A. & Grafen, A. (2005). The illusion of invariant quantities in life histories. *Science,* 309, 1236–1239.

Neel, J.V. (1962). Diabetes mellitus: a "thrifty" genotype rendered detrimental by "progress". *American Journal of Human Genetics,* 14, 353–362.

Nei, M. (1972). Genetic distance between populations. *American Naturalist,* 106, 283–292.

(1973) Analysis of genetic diversity in subdivided populations. *Proceedings of the National. Academy of Sciences of the United States of America,* 70, 3321–3323.

Newman, T.K., Fairbanks, L.A., Pollack, D. & Rogers, J. (2002). Effectiveness of human microsatellite loci for assessing paternity in a captive colony of vervets (*Chlorocebus aethiops sabaeus*). *American Journal of Primatology*, 56(4), 237–243.

Niu, Y., Shen, B., Cui, Y. *et al.* (2014). Generation of gene-modified cynomolgus monkey via Cas9/RNA-mediated gene targeting in one-cell embryos. *Cell*, 156, 836–843.

Noe, R. & Laporte, M. (2014). Socio-spatial cognition in vervet monkeys. *Animal Cognition*, 17, 597–607.

Novak, M.A. (2003). Self-injurious behavior in rhesus monkeys: new insights into its etiology, physiology, and treatment. *American Journal of Primatology*, 59(1), 3–19.

Nowack, J., Mzilikazi, N. & Dausmann, K.H. (2010). Torpor on demand: heterothermy in the non-lemur primate Galago moholi. *PLoS One*, 5, e10797.

Nozawa, K., Shotake, T., Kawamoto, Y. & Tanabe, Y. (1982). Population genetics of Japanese monkeys, II. Blood protein polymorphisms and population structure. *Primates*, 23 (2), 252–271.

Nozawa, K., Shotake, T., Minezawa, M., Kawamoto, Y., Hayasaka, K., Kawamoto, S. & Ito, S. (1991). Population genetics of Japanese monkeys, III. Ancestry and differentiation of local populations. *Primates*, 32(4), 411–435.

Nunn, C.L. (1999). The number of males in primate social groups: a comparative test of the sociological model. *Behavioral Ecology and Sociobiology*, 46(1), 1–13.

Nunn, C.L., Thrall, P.H., Leendertz, F.H. & Boesch, C. (2011). The spread of fecally transmitted parasites in socially-structured populations. *PLoS One*, 6(6), e21677.

Oates, J.F. (1988). The distribution of *Cercopithecus* monkeys in West African forests. In A. Gautier-Hion, F. Bourliere, J.P. Gautier-Hion & J. Kingdon, eds., *A Primate Radiation: Evolutionary Biology of the African Guenons*. Cambridge: Cambridge University Press, pp. 79–103.

Ober, C., Olivier, T.J. & Buettner-Janusch, J. (1978). Carbonic anhydrase heterozygosity and FST distributions in Kenyan baboon troops. *American Journal of Physical Anthropology*, 48, 95–100.

(1980). Genetic aspects of migration in a rhesus monkey population. *Journal of Human Evolution*, 9, 187–203.

Obregon-Tito, A.J., Tito, R.Y., Metcalf, J.L. et al. (2015). Subsistence strategies in traditional societies distinguish gut microbiomes. *Nature Communications*, 6, 6505.

O'Brien, S.J. & Mayr, E. (1991). Bureaucratic mischief: recognizing endangered species and subspecies. *Science*, 251(4998), 1187–1188.

Ockerse, T. (1959). The anatomy of the teeth of the vervet monkey. *Journal of the Dental Association of South Africa*, 14, 209–226.

Ogawa, H. & Takahashi, H. (2003). Triadic positions of Tibetan macaques huddling at a sleeping site. *International Journal of Primatology*, 24, 591–606.

Okoye, A., Meier-Schellersheim, M., Brenchley, J.M. et al. (2007). Progressive CD4$^+$ central memory T cell decline results in CD4$^+$ effector memory insufficiency and overt disease in chronic SIV infection. *Journal of Experimental Medicine*, 204, 2171–2185.

Olivier, T.J., Buettner-Janusch, J. & Buettner-Janusch, V. (1974). Carbonic anhydrase isoenzymes in nine troops of Kenya baboons, *Papio cynocephalus* (Linnaeus 1776). *American Journal of Physical Anthropology*, 41, 175–189.

Olivier, T.J., Coppenhaver, D.H. & Steinberg, A.G. (1986). Distribution of immunoglobulin allotypes among local populations of Kenya olive baboons. *American Journal of Physical Anthropology*, 70(1), 29–38.

Olupot, W. & Waser, P.M. (2001). Activity patterns, habitat use and mortality risks of mangabey males living outside social groups. *Animal Behavior*, 61, 227–1235.

Ostner, J. & Schülke, O. (2018). Linking sociality to fitness in primates: a call for mechanisms. *Advances in the Study of Behavior*, 50, 127–175.

Otsyula, M.G., Gettie, A., Suleman, M., Tarara, R., Mohamed, I. & Marx, P. (1995). Apparent lack of vertical transmission of simian immunodeficiency virus (SIV) in naturally infected African green monkeys, *Cercopithecus aethiops*. *Annals of Tropical Medicine and Parasitology*, 89, 573–576.

Ou, J., Carbonero, F., Zoetendal, E.G., DeLaney, J.P., Wang, M., Newton, K., Gaskins, H.R. & O'Keefe, S.J.D. (2013). Diet, microbiota, and microbial metabolites in colon cancer risk in rural Africans and African–Americans. *American Journal of Clinical Nutrition*, 98(1), 111–120.

Pahl, L., Schubert, S., Skawran, B., Sandbothe, M., Schmidtke, J. & Stuhrmann, M. (2013). 1,25-Dihydroxyvitamin D decreases HTRA1 promoter activity in the rhesus monkey – a plausible explanation for the influence of vitamin D on age-related macular degeneration? *Experimental Eye Research*, 116, 234–239.

Paiardini, M., Cervasi B., Reyes-Aviles, E. et al. (2011). Low levels of SIV infection in sooty mangabey central memory CD4$^+$ T cells are associated with limited CCR5 expression. *Nature Medicine*, 17, 830–836.

Palesch, D., Bosinger, S.E., Tharp, G.K. et al. (2018). Sooty mangabey genome sequence provides insight into AIDS resistance in a natural SIV host. *Nature*, 553(7686), 77–81.

Pampush, J.D. (2010). Human food access and its effects on South African vervet body mass. MSc thesis. Milwaukee, WI: University of Wisconsin-Milwaukee.

Pandrea, I. & Apetrei, C. (2010). Where the wild things are, pathogenesis of SIV infection in African nonhuman primate hosts. *Current HIV/AIDS Report*, 7, 28–36.

Pandrea, I., Apetrei, C., Dufour, J. et al. (2006). Simian immunodeficiency virus (SIV) SIVagm.sab infection of Caribbean African green monkeys: new model of the study of SIV pathogenesis in natural hosts. *Journal of Virology*, 80, 4858–4867.

Pandrea, I., Apetrei, C., Gordon, S. et al. (2007). Paucity of CD4$^+$CCR5$^+$ T cells is a typical feature of natural SIV hosts. *Blood*, 109, 1069–1076.

Pandrea, I., Gaufin, T., Brenchley, J.M. et al. (2008). Cutting edge, experimentally induced immune activation in natural hosts of simian immunodeficiency virus induces significant increases in viral replication and CD4$^+$ T cell depletion. *Journal of Immunology*, 181, 6687–6691.

Pandrea, I.V., Gautam, R., Ribeiro, R.M. et al. (2007). Acute loss of intestinal CD4$^+$ T cells is not predictive of simian immunodeficiency virus virulence. *Journal of Immunology*, 179, 3035–3046.

Pandrea, I., Kornfeld, C., Ploquin, M.J.I. et al. (2005). Impact of viral factors on very early in vivo replication profiles in SIVagm-infected African green monkeys. *Journal of Virology*, 79, 6249–6259.

Pandrea, I., Onanga, R., Rouquet, P., Bourry, O., Ngari, P., Wickings, E.J., Roques, P. & Apetrei, C. (2001). Chronic SIV infection ultimately causes immunodeficiency in African non-human primates. *AIDS*, 15, 2461–2462.

Pandrea, I., Parrish, N.F., Raehtz, K. et al. (2012). Mucosal simian immunodeficiency virus transmission in African green monkeys, susceptibility to infection is proportional to target cell availability at mucosal sites. *Journal of Virology*, 86, 4158–4168.

Pandrea, I., Ribeiro, R.M., Gautam, R. et al. (2008). Simian immunodeficiency virus SIVagm dynamics in African green monkeys. *Journal of Virology*, 82, 3713–3724.

Pandrea, I., Silvestri, G. & Apetrei, C. (2009). AIDS in African nonhuman primate hosts of SIVs: a new paradigm of SIV infection. *Current HIV Research*, 6, 57–72.

Pandrea, I., Sodora, D.L., Silvestri, G. & Apetrei, C. (2008). Into the wild, simian immunodeficiency virus (SIV) infection in natural hosts. *Trends in Immunology*, 29, 419–428.

Parker, K.J., Rainwater, K.L., Buckmaster, C.L., Schatzberg, A.F., Lindley, S.E. & Lyons, D.M. (2007). Early life stress and novelty seeking behavior in adolescent monkeys. *Psychoneuroendocrinology*, 32, 785–92.

Pasternak, H., Brown, L.R., Kienzle, S., Fuller, A., Barrette, L. & Henzi, S.P. (2013). Population ecology of vervet monkeys in a high latitude, semi-arid riparian woodland. *Koedoe*, 55(1), 1078–1086.

Pawlak, C.R., Ho, Y.J., & Schwarting, R.K. (2008). Animal models of human psychopathology based on individual differences in novelty-seeking and anxiety. *Neuroscience & Biobehavioral Reviews*, 32(8), 1544–1568.

Pedersen, N., Lowenstine, L., Marx, P. et al. (1986). The causes of false-positives encountered during the screening of old-world primates for antibodies to human and simian retroviruses by ELISA. *Journal of Virological Methods*, 14, 213–228.

Pedersen, A.B., Altizer, S., Poss, M., Cunningham, A.A. & Nunn, C.L. (2005). Patterns of host specificity and transmission among parasites of wild primates. *International Journal of Parasitology*, 35(6), 647–657.

Perry, J.R., Day, F. & Elks, C.E. (2014). Parent-of-origin-specific allelic associations among 106 genomic loci for age at menarche. *Nature*, 514, 92–97.

Perry, G.H., Marioni, J.C., Melsted, P. & Gilad, Y. (2010). Genomic-scale capture and sequencing of endogenous DNA from feces. *Molecular Ecology*, 19, 5332–5344.

Petrželková, K.J., Hasegawa, H., Appleton, C.C., Huffman, M.A., Archer, C.E., Moscovice, L.R., Mapua, M.I., Singh, J. & Kaur, T. (2010). Gastrointestinal parasites of indigenous and introduced primate species of Rubondo Island National Park, Tanzania. *International Journal of Primatology*, 31(5), 920–936.

Phillips-Conroy, J.E. & Jolly, C.J. (1988). Dental eruption schedules of wild and captive baboons. *American Journal of Primatology*, 15(1), 17–29.

Philips-Conroy, J.E., Jolly, C.J., Nystrom, P. & Hemmalin, H.A. (1992). Migration of male hamadryas baboons into anubis groups in the Awash National Park, Ethiopia. *International Journal of Primatology*, 13(4), 455–476.

Phillips-Conroy, J.E., Jolly, C.J., Petros, B., Allan, J.S. & Desrosiers, R.C. (1994). Sexual transmission of SIVagm in wild grivet monkeys. *Journal of Medical Primatology*, 23, 1–7.

Pickford, M. (1988). Habitat and locomotion in Miocene cercopithecoids. In A. Gautier-Hion, F. Bourliere, J.P. Gautier-Hion & J. Kingdon, eds., *A Primate Radiation: Evolutionary Biology of the African Guenons*. Cambridge: Cambridge University Press, pp. 35–53.

Pocock, R.I. (1907). A monographic revision of the monkeys of the genus *Cercopithecus*. *Proceedings Zoological Society London*, 1, 677–746.

Poirier, F.E. (1972). The St Kitts green monkeys (*Cercopithecus aethiops sabaeus*): ecology, population dynamics, and selected behavioral traits. *Folia Primatologica*, 17, 20–55.

Ponsa, M., Egozcue, J. & Garcia, M. (1994). The phylogeny of guenons using PAUP analysis of cytogenetic characters. In B. Thierry, J.R. Anderson, J.J. Roeder & N. Herrenschmidt, eds., *Current Primatology, Vol. I, Ecology and Evolution.* Strasbourg, France: University Louis Pasteur, pp. 387–392.

Popp, J.L. (1983). Ecological determinism in the life histories of baboons. *Primates,* 24(2), 198–210.

Potts, R. (1998). Variability selection in hominid evolution. *Evolutionary Anthropology: Issues, News, and Reviews,* 7, 81–96.

Prado-Martinez, J., Sudman, P.H. & Kidd, J.M. (2013). Great ape genetic diversity and population history. *Nature,* 499 (7459), 471–5.

Prentice, A.M., Hennig, B.J. & Fulford, A.J. (2008). Evolutionary origins of the obesity epidemic: natural selection of thrifty genes or genetic drift following predation release? *International Journal of Obesity,* 32, 1607–1610.

Promislow, D.E.L. & Harvey, P.H. (1990). Living fast and dying young: a comparative analysis of life-history variation among mammals. *Journal of Zoology,* 220(3), 417–437.

Pruetz, J.D. & Bertolani, P. (2007). Savanna chimpanzees, *Pan troglodytes verus,* hunt with tools. *Current Biology,* 17(5), 412–417.

(2009). Chimpanzee (*Pan troglodytes verus*) behavioral responses to stresses associated with living in a savanna–mosaic environment, implications for hominin adaptations to open habitats. *PaleoAnthropology,* 252–262.

Pruetz, J.D. & Isbell, L.A. (2000). Correlations of food distribution and patch size with agonistic interactions in female vervets (*Chlorocebus aethiops*) and patas monkeys (*Erythrocebus patas*) living in simple habitats. *Behavioral Ecology and Sociobiology,* 49, 38–47.

Prum, R.O. & Torres, R.H. (2004). Structural colouration of mammalian skin: convergent evolution of coherently scattering dermal collagen arrays. *Journal of Experimental Biology,* 207, 2157–2172.

Pulliam, H.R. (1973). On the advantages of flocking. *Journal of Theoretical Biology,* 38, 419–422.

Pusey, A.E., Oehlert, G.W., Williams, J.M. & Goodall, J. (2005). Influence of ecological and social factors on body mass of wild chimpanzees. *International Journal of Primatology,* 26(1), 3–31.

Pusey, A. & Packer, C. (1987). Dispersal and philopatry. In B. Smuts, D. Cheney, R. Seyfarth, R. Wrangham & T. Struhsaker, eds., *Primate Societies.* Chicago, IL: University of Chicago Press, pp. 250–266.

Raaum, R.L., Sterner, K.N., Noviello, C.M., Stewart, C.B. & Disotell, T.R. (2005). Catarrhine primate divergence dates estimated from complete mitochondrial genomes: concordance with fossil and nuclear DNA evidence. *Journal of Human Evolution,* 48(3), 237–257.

Rainwater, D.L., Cox, L.A., Rogers, J., VandeBerg, J.L. & Mahaney, M.C. (2009). Localization of multiple pleiotropic genes for lipoprotein metabolism in baboons. *Journal of Lipid Research*, 50, 1420–1428.

Raleigh, M.J. & McGuire, M.T. (1989). Female influences on male dominance in captive vervet monkeys, *Cercopithecus aethiops sabaeus*. *Animal Behavior*, 38, 59–67.

(1990). Social influences on endocrine function in male vervet monkeys. In T.E. Ziegler & F.B. Bercovitch, eds., *Socioendocrinology of Primate Reproduction*. New York: Wiley-Liss, pp. 95–111.

Rapkin, A.J., Pollack, D.B., Raleigh, M.J., Stone, B. & McGuire, M.T. (1995). Menstrual cycle and social behavior in vervet monkeys. *Psychoneuroendocrinology*, 20(3), 289–297.

Ravel, J., Gajer, P., Abdo, Z. *et al.* (2011). Vaginal microbiome of reproductive-age women. *Proceedings of the National Academy of Science of the United States of America*, 108, 4680–4687.

Ravosa, M.J. & Hylander, W.L. (1993). Relative growth of the limbs and trunk in sifakas: heterochronic, ecological, and functional considerations. *American Journal of Physical Anthropology*, 92(4), 499–520.

Raz, A., Vaadia, E. & Bergman, H. (2000). Firing patterns and correlations of spontaneous discharge of pallidal neurons in the normal and the tremulous 1-methyl-4-phenyl-1,2,3,6-tetrahydropyridine vervet model of parkinsonism. *Journal of Neuroscience*, 20, 8559–8571.

Read, A.F. & Harvey, P.H. (1989). Life history differences among the eutherian radiations. *Journal of Zoology*, 219(2), 329–353.

Reif, A. & Lesch, K.P. (2003). Toward a molecular architecture of personality. *Behavioral Brain Research*, 139, 1–20.

Reimers, M., Schwarzenberger, F. & Preuschoft, S. (2007). Rehabilitation of research chimpanzees: stress and coping after long-term isolation. *Hormones and Behavior*, 51(3), 428–435.

Reinhardt, V. & Garza-Schmidt, M. (2000). Daily feeding enrichment for laboratory macaques: inexpensive options. *Laboratory Primate Newsletter*, 39 (2), 8–10.

Renevey, N., Bshary, R. & van de Waal, E. (2013). Philopatric vervet monkey females are the focus of social attention rather independently of rank. *Behaviour*, 150, 599–615.

Rensch, B. (1950). Die abhangigkeit der relativen sexualdifferenz von der körpergroße. *Bonner Zoologische Beiträge*, 1, 58–69.

Reznick, D., Bryant, M.J. & Bashey, F. (2002). r- and K-selection revisited: the role of population regulation in life-history evolution. *Ecology*, 83, 1509–1520.

Reznick, D., Nunney, L. & Tessler, A. (2000). Big houses, big cars, superfleas and the costs of reproduction. *TREE*, 15(10), 421–425.

Richard, A.F. (1985). *Primates in Nature*. New York: W.H. Freeman and Company.

Richard, A.F., Dewar, R.E., Schwartz, M. & Ratsirarson, J. (2000). Mass change, environmental variability and female fertility in wild *Propithecus verrauxi*. *Journal of Human Evolution*, 39(4), 381–391.

Richerson, P. & Boyd, R. (2005). *Not by Genes Alone*. Chicago: Chicago University Press.

Riddick, N.E., Wu, F., Matsuda, K. et al. (2015). Simian immunodeficiency virus SIVagm efficiently utilizes non-CCR5 entry pathways in African green monkey lymphocytes, potential role for GPR15 and CXCR6 as viral coreceptors. *Journal of Virology*, 90, 2316–2331.

Riley, E.P. (2006). Ethnoprimatology: Towards reconciliation of biological and cultural anthropology. *Ecological and Environmental Anthropology*, 2(2), 1–10.

(2007). The human–macaque interface: conservation implications of current and future overlap and conflict in Lore Lindu National Park, Sulawesi, Indonesia. *American Anthropologist*, 109, 473–484.

(2010). The importance of human–macaque folklore for conservation in Lore National Park, Sulawesi, Indonesia. *Oryx*, 44(2), 235–240.

Riley, E.P. & Ellwanger, A.L. (2013). Methods in ethnoprimatology, exploring the human–non-human primate interface. In E.J. Sterling, N. Bynum & M.E. Blair, eds., *Primate Ecology and Conservation, A Handbook of Techniques*. Oxford, UK: Oxford University Press, pp. 128–150.

Riley, E.P. & Priston, N.E. (2010). Macaques in farms and folklore: exploring the human–nonhuman primate interface in Sulawesi, Indonesia. *American Journal of Primatology*, 72(10), 848–854.

Rodman, P.S. (1988). Resources and group sizes in primates. In C.N. Slobodchikoff, ed., *The Ecology of Social Behavior*. New York: Academic Press, pp. 83–108.

Rodríguez, R.L. & Al-Wathiqui, N. (2012). Causes of variation in sexual allometry: a case study with the mating signals and genitalia of Enchenopa treehoppers (*Hemiptera membracidae*). *Ethology Ecology & Evolution*, 24(2), 187–197.

Rodríguez, R.L., Araya-Salas, M., Gray, D.A., Reichert, M.S., Symes, L.B., Wilkins, M.R., Safran, R.J. & Höbel, G. (2015a). How acoustic signals scale with individual body size, common trends across diverse taxa. *Behavioral Ecology*, 26, 168–177.

Rodríguez, R. L., Cramer, J.D., Schmitt, C.A. et al. (2015b). Adult age confounds estimates of static allometric slopes in a vertebrate. *Ethology, Ecology & Evolution*, 27, 412–431.

Rodríguez, R.L., Cramer, J.D., Schmitt, C.A., Gaetano, T.J., Grobler, J.P., Freimer, N.B. & Turner, T.R. (2015c). The static allometry of sexual and non-sexual traits in vervet monkeys. *Biological Journal of the Linnean Society*, 114, 527–537.

Roederer, M. (1995). T-cell dynamics of immunodeficiency. *Nature Medicine*, 1, 621–622.

Roff, D.A. (2002). *Life History Evolution*. Sunderland, MA: Sinauer Associates, Inc.

Rogers, J.A. (1989). Genetic structure and microevolution in a population of Tanzanian yellow baboons (Papio hamadryas cynocephalus). PhD thesis. New Haven, CT: Yale University.

Rogers, J. (2000). Molecular genetic variation and population structure in *Papio* baboons. In P.F. Whitehead & C.J. Jolly, eds., *Old World Monkeys*. Cambridge: Cambridge University Press, pp. 57–76.

Rogers, J. & Gibbs, R.A. (2014). Comparative primate genomics: emerging patterns of genome content and dynamics. *Nature Reviews Genetics*, 15, 347–359.

Rogers, J. & Kidd, K.K. (1993). Nuclear DNA polymorphisms in a wild population of yellow baboons (*Papio hamadryas cynocephalus*) from Mikumi National Park, Tanzania. *American Journal of Physical Anthropology*, 90(4), 477–486.

Romero, R., Hassan, S.S., Gajer, P. et al. (2014). The composition and stability of the vaginal microbiota of normal pregnant women is different from that of non-pregnant women. *Microbiome*, 2, 4.

Rose, M.R. (1985). Life history evolution with antagonistic pleiotropy and overlapping generations. *Theoretical Population Biology*, 28(3), 342–358.

Rothman, J.M., Chapman, C.A., Twinomugisha, D., Wasserman, M.D., Lambert, J.E. & Goldberg, T. (2008). Measuring physical traits of primates remotely, the use of parallel lasers. *American Journal of Primatology*, 70, 1191–1195.

Rowe, L. & Houle, D. (1996). The lek paradox and the capture of genetic variance by condition dependent traits. *Proceedings of the Royal Society of London B: Biological Sciences*, 263(1375), 1415–1421.

Rowe, N. (1996). *The Pictorial Guide to the Living Primates*. Charlestown, RI: Pogonias Press.

Rowell, T.E. (1988). The social system of guenons, compared with baboons, macaques and mangabeys. In A. Gautier-Hin, F. Bourliere, J.P. Gautier & J. Kingdon, eds., *A Primate Radiation: Evolutionary Biology of the African Guenons*. Cambridge: Cambridge University Press, pp. 439–451.

Rowell, T.E. & Olson, D.K. (1983). Alternative mechanisms of social organization in monkeys. *Behaviour*, 86, 31–54.

Rowell, T.E. & Richards, S.M. (1979). Reproductive strategies of some African monkeys. *Journal of Mammalogy*, 60, 58–69.

Rukstalis, M. & French, J.A. (2004). Exposure to conspecific vocalizations modulates stress responses in marmosets (*Callithrix kuhlii*). *American Journal of Primatology*, 60(S1), 129–130.

(2004). Conspecific vocal signals moderate urinary cortisol excretion in isolated marmosets (*Callithrix kuhlii*). *American Journal of Primatology*, 61(S1), 80.

Rutenberg, G.W., Coelho Jr., A.M., Lewis, D.S., Carey, K.D. & McGill Jr., H.C. (1987). Body composition in baboons, evaluating a morphometric method. *American Journal of Primatology*, 12, 275–285.

Ruvolo, M. (1988). Genetic evolution in the African guenons. In A. Gautier-Hion, F. Bourliere, J.P. Gautier-Hion & J. Kingdon, eds., *A Primate Radiation: Evolutionary Biology of the African Guenons*. Cambridge: Cambridge University Press, pp. 127–139.

Ryan, S. J., Brashares, J. S., Walsh, C., Milbers, K., Kilroy, C. & Chapman, C. A. (2012). A survey of gastrointestinal parasites of olive baboons (*Papio anubis*) in human settlement areas of Mole National Park, Ghana. *Journal of Parasitology*, 98(4), 885–888.

Sackett, G.P., Bowman, R.E., Meyer, J.S., Tripp, R.L. & Grady, S.S. (1973). Adrenocortical and behavioral reactions by differentially raised rhesus monkeys. *Physiological Psychology*, 1, 209–212.

Sade, D.S. & Hildrech, R.W. (1965). Notes on the green monkey (*Cercopithecus aethiops sabaeus*) on St Kitts, West Indies. *Caribbean Journal of Science*, 5, 67–81.

Sadlier, R.M.F.S. (1969). *The Ecology of Reproduction in Wild and Domestic Mammals*. London, UK: Methuen & Co Ltd.

Saj, T.L. & Sicotte, P. (2007). Predicting the competitive regime of female *Colobus vellerosus* from the distribution of food resources. *International Journal of Primatology*, 28, 315–336.

Santiago, M.L., Bibollet-Ruche, F., Gross-Camp, N. et al. (2003). Noninvasive detection of simian immunodeficiency virus infection in a wild-living l'hoest's monkey (*Cercopithecus lhoesti*). *AIDS Research and Human Retroviruses*, 19, 1163–1166.

Santiago, M.L., Range, F., Keele, B.F. et al. (2005). Simian immunodeficiency virus infection in free-ranging sooty mangabeys (*Cercocebus atys atys*) from the Tai Forest, Cote d'Ivoire, implications for the origin of epidemic human immuno-deficiency virus type 2. *Journal of Virology*, 79, 12515–12527.

Santiago, M.L., Rodenburg, C.M., Kamenya, S. et al. (2002). SIVcpz in wild chimpanzees. *Science*, 295, 465.

Santos, L.R., Pearson, H.M., Spaepen, G.M., Tsao, F. & Hauser, M.D. (2006). Probing the limits of tool competence: experiments with two non-tool-using species (*Cercopithecus aethiops* and *Saguinus oedipus*). *Animal Cognition*, 9, 94–109.

Sapolsky, R.M. (2005). The influence of social hierarchy on primate health. *Science*, 308, 648–652.

Sargis, E.J., Terranova, C.J. & Gebo, D.L. (2008). Evolutionary morphology of the guenon postcranium and its taxonomic implication. In E.J. Sargis & M. Dagosto, eds., *Mammalian Evolutionary Morphology*. Dordrecht, The Netherlands: Springer, pp. 361–372.

Schillaci, M.A., Castellini, M.J., Stricker, C.A. et al. (2014). Variation in hair $\delta^{13}C$ and $\delta^{15}N$ values in long-tailed macaques (*Macaca fascicularis*) from Singapore. *Primates*, 55, 25–34.

Schillaci, M.A., Jones-Engel, L., Lee, B.P.Y.H. et al. (2007). Morphology and somatometric growth of long-tailed macaques (*Macaca fascicularis fascicularis*) in Singapore. *Biological Journal of the Linnean Society*, 92, 675–694.

Schillaci, M.A. & Stallmann, R.R. (2005). Ontogeny and sexual dimorphism in booted macaques (*Macaca ochreata*). *Journal of Zoology*, 267, 19–29.

Schmitt, C.A., Service, S.K., Jasinska, A.J., Dyer, T.D., Jorgensen, M.J., Cantor, R.M., Weinstock, G.M., Blangero, J., Kaplan, J.R. & Freimer, N.B. (2018). Obesity and obesogenic growth are both highly heritable and modified by diet in a nonhuman primate model, the African green monkey (*Chlorocebus aethiops sabaeus*). *International Journal of Obesity*, 42(4), 765–774.

Schnorr, S.L., Candela, M., Rampelli, S. et al. (2014). Gut microbiome of Hadza hunter–gatherers. *Nature Communications*, 5, 3654.

Schoene, C.U.R. & Brend, S.A. (2002). Primate sanctuaries – a delicate conservation approach. *South African Journal of Wildlife Research*, 32(2), 109–113.

Schoeninger, M.J. & DeNiro, M.J. (1984). Nitrogen and carbon isotopic composition of bone collagen from marine and terrestrial animals. *Geochimica et Cosmochimica Acta*, 48, 625–639.

Schoeninger, M.J., Moore, J. & Sept, J.M. (1999). Subsistence strategies of two "savanna" chimpanzee populations, the stable isotope evidence. *American Journal of Primatology*, 49, 297–314.

Schuett, W., Tregenza, T. & Dall, S.R.X. (2010). Sexual selection and animal personality. *Biological Reviews*, 85, 217–246.

Schülke, O., Bhagavatula, J., Vigilant, L. & Ostner, J. (2010). Social bonds enhance reproductive success in male macaques. *Current Biology*, 20, 2207–2210.

Schulte-Hostedde, A.I., Kuula, S., Martin, C., Schank, C.C.M. & Lesbarrères, D. (2011). Allometry and sexually dimorphic traits in male anurans. *Journal of Evolutionary Biology*, 24, 1154–1159.

Schulte-Hostedde, A.I., Zinner, B., Millar, J.S. & Hickling, G.J. (2005). Restitution of mass–size residuals: validating body condition indices. *Ecology*, 86(1), 155–163.

Schurr, M.R., Fuentes, A., Luecke, E., Cortes, J. & Shaw, E. (2012). Intergroup variation in stable isotope ratios reflects anthropogenic impact on the Barbary macaques (*Macaca sylvanus*) of Gibraltar. *Primates*, 53, 31–40.

Schwarz, E. (1926). Die meerkatzen der *Cercopithecus aethiops*-gruppe. *Zeitschrift für Säugetierkunde*, 1, 28–47.

Schwartz, S.M. & Kemnitz, J.W. (1992). Age- and gender-related changes in body size, adiposity, and endocrine and metabolic parameters in free-ranging rhesus macaques. *American Journal of Physical Anthropology*, 89(1), 109–121.

Schwartz, S.M., Wilson, M.E., Walker, M.L. & Collins, D.C. (1988). Dietary influences on growth and sexual maturation in premenarchial rhesus monkeys. *Hormones and Behavior*, 22(2), 231–251.

Selander, R.K. & Kaufman, D.W. (1973). Genic variability and strategies of adapatation in animals. *Proceedings of the National Academy of Sciences of the United States of America*, 70(6), 1875–1877.

Semple, S. & Higham, J.P. (2013). Primate signals: current issues and perspectives. *American Journal of Primatology*, 75(7), 613–620.

Šešelj, M. (2013). Relationship between dental development and skeletal growth in modern humans and its implications for interpreting ontogeny in fossil hominins. *American Journal of Physical Anthropology*, 150, 38–47.

Setchell, J.M., Charpentier, M.J.E., Bedjabaga, I.B., Reed, P., Wickings, E.J. & Knapp, L.A. (2006). Secondary sexual characters and female quality in primates. *Behavioral Ecology and Sociobiology*, 61, 305–315.

Setchell, J.M., Lee, P.C., Wickings, E.J. & Dixon, A.F. (2001). Growth and ontogeny of sexual size dimorphism in the mandrill (*Mandrillus sphinx*). *American Journal of Physical Anthropology*, 115, 349–60.

Seyfarth, R.M. (1977). A model of social grooming among adult female monkeys. *Journal of Theoretical Biology*, 65, 671–698.

(1980). The distribution of grooming and related behaviors among adult female vervet monkeys. *Animal Behavior*, 28, 798–813.

Seyfarth, R.M., Cheney, D.L. & Marler, P. (1980a). Monkey responses to three different alarm calls, evidence of predator classification and semantic communication. *Science*, 210, 801–803.

(1980b). Vervet monkey alarm calls: Semantic communication in a free-ranging primate. *Animal Behaviour*, 28, 1070–1094.

Seyfarth, R.M. & Cheney, D.L. (1984). Grooming, alliances and reciprocal altruism in vervet monkeys. *Nature*, 308(5959), 541–543.

Shannon, C., Champoux, M. & Suomi, S.J. (1998). Rearing condition and plasma cortisol in rhesus monkey infants. *American Journal of Primatology*, 46(4), 311–321.

Shannon, C., Schwandt, M.L., Champoux, M., Shoaf, S.E., Suomi, S.J., Linnoila, M. & Higley, J.D. (2005). Maternal absence and stability of individual differences in CSF 5-HIAA concentrations in rhesus monkey infants. *American Journal of Psychiatry*, 162, 1658–1664.

Sharp, P.M. & Hahn, B.H. (2011). Origins of HIV and the AIDS pandemic. *Cold Spring Harbor Perspectives in Medicine*, 1, a006841.

Sharp, P.M., Bailes, E., Gao, F., Beer, B.E., Hirsch, V.M. & Hahn, B.H. (2000). Origins and evolution of AIDS viruses, estimating the time-scale. *Biochemical Society Transactions*, 28, 275–282.

Shea, B.T. (1986). Ontogenetic approaches to sexual dimorphism in anthropoids. *Human Evolution*, 1, 97–110.

Shelley, E.L. & Blumstein, D.T. (2005). The evolution of vocal alarm communication in rodents. *Behavioral Ecology*, 16, 169–177.

Shimada, M.K. (2000). Geographic distribution of mitochondrial DNA variations among grivet (*Cercopithecus aethiops aethiops*) populations in central Ethiopia. *International Journal of Primatology*, 21(1), 113–129.

Shimada, M.K. & Shotake, T. (1997). Genetic variation of blood proteins within and between local populations of grivet monkey (*Cercopithecus aethiops aethiops*) in central Ethiopia. *Primates*, 38(4), 399–414.

Shimada, M.K., Terao, K. & Shotake, T. (2002). Mitochondrial sequence diversity within a subspecies of savanna monkeys (*Cercopithecus aethiops*) is similar to that between subspecies. *Journal of Heredity*, 93(1), 9–18.

Shine, R. (1988). The evolution of large body size in females: a critique of Darwin's "fecundity advantage" model. *American Naturalist*, 131, 124–131.

Shingleton, A.W., Frankino, W.A., Flatt, T., Nijhout, H.F. & Emlen, D.J. (2007). Size and shape, the developmental regulation of static allometry in insects. *BioEssays*, 29, 536–548.

Shotake, T. & Nozawa, K. (1984). Blood protein variations in baboons. II. Genetic variability within and among herds of gelada baboons in the central Ethiopian plateau. *Journal of Human Evolution*, 13(3), 265–274.

Shultz, S., Noë, R., McGraw, W.S. & Dunbar, R.I.M. (2004). A community-level evaluation of the impact of prey behavioural and ecological characteristics on predator diet composition. *Proceedings of the Royal Society London B*, 271, 725–732.

Sigg, H.H., Stolba, A., Abegglen, J.J. & Dasser, V. (1982). Life history of hamadryas baboons: physical development, infant mortality, reproductive parameters and family relationships. *Primates*, 23, 473–487.

Sih, A. & Bell, A.M. (2008). Insights for behavioral ecology from behavioral syndromes. *Advances in the Study of Behavior*, 38, 227–281.

Silk, J.B. (2007a). Social components of fitness in primate groups. *Science*, 317, 1347–1351.

(2007b). The adaptive value of sociality in mammalian groups. *Philosophical Transactions of the Royal Society B, Biological Sciences*, 362, 539–599.

Silk, J.B., Alberts, S.C. & Altmann, J. (2003). Social bonds of female baboons enhance infant survival. *Science*, 302, 1231–1234.

Silk, J.B., Beehner, J.C., Bergman, T.J., Crockford, C., Engh, A.L., Moscovice, L.R., Wittig, R.M., Seyfarth, R.M. & Cheney, D.L. (2009). The benefits of social capital: close social bonds among female baboons enhance offspring survival. *Proceedings of the Royal Society B*, 276, 3099–3104.

Smith, C.C. (1978). Feeding behaviour and social organization in howling monkeys. In T.H. Clutton-Brock, ed., *Primate Ecology: Studies of Feeding and Ranging Behaiour in Lemurs, Monkeys and Apes*. New York, NY: Academic Press, pp. 97–126.

Smith, C.C. & Fretwell, S.D. (1974). The optimal balance between size and number of offspring. *The American Naturalist*, 108(962), 499–506.

Smith, B.R. & Blumstein, D.T. (2008). Fitness consequences of personality: a meta analysis. *Behavioral Ecology*, 19(2), 448–455.

Smuts, B.B. & Smuts, R.W. (1993). Male aggression and sexual coercion of females in nonhuman primates and other mammals: evidence and theoretical implications. *Advances in the Study of Behavior*, 22, 1–63.

Smuts, B.B., Cheney, D.L., Seyfarth, R.M. & Wrangham, R.W. (1987). *Primate Societies*. Chicago: University of Chicago Press.

Sokal, R.R. & Rohlf, F.J. (1981). *Biometry: The Principles and Practice of Statistics in Biological Research*, 2nd edn. New York: W.H. Freeman and Company.

Spitze, K. (1991). *Chaoborus* predation and life-history evolution in *Daphnia pulex*: temporal pattern of population diversity, fitness, and mean life history. *Evolution*, 45, 82–92.

Sponsel, L.E. (1997). The human niche in Amazonia, explorations in ethnoprimatology. In W.G. Kinzey, ed., *New World Primates*. New York: Holt, Rinehart and Winston, pp. 111–159.

Sponheimer, M., Loudon, J.E., Codron, D., Howells, M.E., Pruetz, J.D., Codron, J., de Ruiter, D.J. & Lee-Thorp, J.A. (2006). Do "savanna" chimpanzees consume C_4 resources? *Journal of Human Evolution*, 51, 128–133.

Sprague, D. (2002). Monkeys in the backyard: encroaching wildlife and rural communities in Japan. In A. Fuentes & L.D. Wolfe, eds., *Primates Face to Face: The Conservation Implications of Human and Nonhuman Primate Interconnections*. Cambridge: Cambridge University Press, pp 254–272.

Sprague, D. & Iwasaki, N. (2006). Coexistence and exclusion between humans and monkeys in Japan: is either really possible? *Ecological and Environmental Anthropology*, 2, 30–43.

Stanford, C.B. (1995). The influence of chimpanzee predation on group size and anti-predator behaviour in red colobus monkeys. *Animal Behaviour*, 49(3), 577–587.

Stearns, S.C. (1992). *The Evolution of Life Histories*. New York: Oxford University Press.

Stearns, S.C., Byars, S.G., Govindaraju, D.R. & Ewbank, D. (2010). Measuring selection in contemporary human populations. *Nature Reviews Genetics*, 11(9), 611.

Sterck, E.H.M., Watts, D.P. & van Schaik, C.P. (1997). The evolution of female social relationships in nonhuman primates. *Behavioral Ecology and Sociobiology*, 41, 291–309.

Stevens, M., Stoddard, M.C. & Higham, J.P. (2009). Studying primate color: towards visual system dependent techniques. *International Journal of Primatology*, 30, 893–917.

Stinson, S. (2012). Growth variation: biological and cultural factors. In S. Stinson, B. Bogin & D. O'Rourke, eds., *Human Biology: An Evolutionary and Biocultural Perspective*, 2nd edn. Hoboken, NJ: Wiley Blackwell, pp. 587–635.

Stoner, K.E. (1996). Prevalence and intensity of intestinal parasites in mantled howling monkeys (*Alouatta palliata*) in northeasten Costa Rica: implications for conservation biology. *Conservation Biology*, 10(2), 539–546.

Strasser, E. & Delson, E. (1987). Cladistic analysis of cercopithecid relationships. *Journal of Human Evolution*, 16(1), 81–99.

Strier, K.B., Altmann, J., Brockman, D.K., Bronikowski, A.M., Cords, M., Fedigan, L.M., Lapp, H., Liu, X., Morris, W.F., Pusey, A.E., Stoinski, T.S. & Alberts, S.C. (2010). The Primate Life History Database, a unique shared ecological data resource. *Methods in Ecology and Evolution*, 1, 199–211.

Struhsaker, T.T. (1967a). Behavior of vervet monkeys (*Cercopithecus aethiops*). *University of California Publications in Zoology*, 82, 1–64.

(1967b). Auditory communication among vervet monkeys (*Cercopithecus aethiops*). In S. Altmann, ed., *Social Communication among Primates*. Chicago: University of Chicago Press, pp. 281–324.

(1967c). Social structure among vervet monkeys (*Cercopithecus aethiops*). *Behaviour*, 29(2), 83–121.

(1967d). Ecology of vervet monkeys (*Cercopithecus aethiops*) in the Masai-Amboseli Game Reserve, Kenya. *Ecology*, 48(6), 891–904.

(1969). Correlates of ecology and social organization among African cercopithecines. *Folia Primatologica*, 11, 80–118.

Strum, S.C. (1991). Weight and age in wild olive baboons. *American Journal of Primatology*, 25(4), 219–237.

Stumpf, R.M., Wilson, B.A., Rivera, A., Yildirim, S., Yeoman, C.J., Polk, J.D., White, B.A. & Leigh, S.R. (2013). The primate vaginal microbiome, comparative context and implications for human health and disease. *American Journal of Physical Anthropology*, 57, 119–134.

Sudmant, P.H., Huddleston, J. & Catacchio, C.R. (2013). Evolution and diversity of copy number variation in the great ape lineage. *Genome Research*, 23, 1373–1382.

Sudmant, P.H., Rausch, T. & Gardner, E.J. (2015). An integrated map of structural variation in 2,504 human genomes. *Nature*, 526, 75–81.

Suomi, S. J. (1982). Animal models of human psychopathology, relevance for clinical psychology. In P. Kendall & J. Butcher, eds., *Handbook of Research Methods in Clinical Psychology*. New York: John Wiley, pp. 249–271.

Suomi, S. (2003). Gene–environment interactions and the neurobiology of social conflict. *Annals of the New York Academy of Sciences*, 1008, 132–139.

Suomi, S.J. (2006). Risk, resilience, and gene × environment interactions in rhesus monkeys. *Annals of the New York Academy of Sciences*, 1094, 52–62.

Svardal, H., Jasinska, A.J., Apetrei, C., Coppola, G., Huang, Y., Schmitt, C.A., Jacquelin, B., Müller-Trutwin, M., Weinstock, G., Grobler, J.P., Wilson, R.K., Turner, T.R., Warren, W.C., Freimer, N.B. & Nordborg, M. (2017). Ancient hybridization and strong adaptation to viruses across African vervet monkey populations. *Nature Genetics*, 49, 1705–1713.

Takahashi, H. (1997). Huddling relationships in night sleeping groups among wild Japanese macaques in Kinkazan Island during winter. *Primates*, 38, 57–68.

Tanner, J.M. (1962). *Growth at Adolescence*, 2nd edn. Oxford, UK: Blackwell.

Tanner, J.M., Wilson, M.E. & Rudman, C.G. (1990). Pubertal growth spurt in the female rhesus monkey: relation to menarche and skeletal maturation. *American Journal of Human Biology*, 2(2), 101–106.

Teaford, M.F. & Ungar, P.S. (2000). Diet and the evolution of the earliest human ancestors. *Proceedings of the National Academy of Science of the United States of America*, 97(25), 13506–13511.

Tecot, S.R., Baden, A.L., Romine, N.K. & Kamilar, J.M. (2012). Infant parking and nesting, not allomaternal care, influence Malagasy primate life histories. *Behavioral Ecology and Sociobiology*, 66(10), 1375–1386.

Teichroeb, J.A., White, M.M.J. & Chapman, C.A. (2015). Vervet monkey (*Chlorocebus pygerythrus*) intragroup spatial positioning: dominants trade-off predation risk for increased food acquisition. *International Journal of Primatology*, 38, 154–176.

Terborgh, J. & Janson, C.H. (1986). The socioecology of primate groups. *Annual Review of Ecology, Evolution, and Systematics*, 17, 111–135.

Thomas, C.D., Cameron, A., Green, R.E. et al. (2004). Extinction risk from climate change. *Nature*, 427, 145–148.

Thompson, C.L., Williams, S.H., Glander, K.E., Teaford, M.F. & Vinyard, C.J. (2014). Body temperature and thermal environment in a generalized arboreal anthropoid, wild mantled howling monkeys (Alouatta palliata). *American Journal of Physical Anthropology*, 154, 1–10.

Thornton, A. & Clutton-Brock, T. (2011). Social learning and the development of individual and group behaviour in mammal societies. *Philosophical Transactions of the Royal Society B*, 366, 978–987.

Toft, II, J.D. & Eberhard, M.L. (1998). Parasitic diseases. In B.T. Bennett, C.R. Abee & R. Henrickson, eds., *Nonhuman Primates in Biomedical Research: Diseases*. San Diego, CA: Academic Press, pp. 111–205.

Tosi, A.J., Buzzard, P.J., Morales, J.C. & Melnick, D.J. (2002). Y-chromosomal window onto the history of terrestrial adaptation in the Cercopithecini. In M.E. Glenn & M. Cords, eds., *The Guenons, Diversity and Adaptation in African Monkeys*. New York, Kluwer Academic Plenum Publishing, pp. 15–26.

Tosi, A.J., Disotell, T.R., Morales, J.C. & Melnick, D.J. (2003). Cercopithecine Y-chromosome data provide a test of competing morphological evolutionary hypotheses. *Molecular Phylogenetics and Evolution*, 27(3), 510–521.

Tosi, A.J., Detwiler, K.M. & Disotell, T.R. (2005). X-chromosomal window into the evolutionary history of the guenons (Primates, Cercopithecini). *Molecular Phylogenetics and Evolution*, 36(1), 58–66.

Tosi, A.J., Melnick, D.J. & Disotell, T.R. (2004). Sex chromosome phylogenetics indicate a single transition to terrestriality in the guenons (tribe Cercopithecini). *Journal of Human Evolution*, 46(2), 223–237.

Traina-Dorge, V., Blanchard, J., Martin, L. & Murphey-Corb, M. (1992). Immunodeficiency and lymphoproliferative disease in an African green monkey dually infected with SIV and STLV-I. *AIDS Research and Human Retroviruses*, 8, 97–100.

Trayford, H.R. & Farmer, K.H. (2013). Putting the spotlight on internally displaced animals (IDAs): a survey of primate sanctuaries in Africa, Asia, and the Americas. *American Journal of Primatology*, 75(2), 116–134.

Trivers, R.L. (1972). Parental investment and sexual selection. In B. Campbell, ed., *Sexual Selection and the Descent of Man, 1871–1971*. Chicago: Aldine, pp. 136–179.

Tung, J., Alberts, S.C. & Wray, G.A. (2010). Evolutionary genetics in wild primates: combining genetic approaches with field studies of natural populations. *Trends in Genetics*, 26(8), 353–362.

Tung, J., Akinyi, M.Y., Mutura, S., Altmann, J., Wray, G.A. & Alberts, S.C. (2011). Allele-specific gene expression in a wild nonhuman primate population. *Molecular Ecology*, 20(4), 725–739.

Tung, J., Barreiro, L.B. & Burns, M.B. (2015a). Social networks predict gut microbiome composition in wild baboons. *eLife*, 4, e05224.

Tung, J., Charpentier, M.J.E., Garfield, D.A., Altmann, J. & Alberts, S.C. (2008). Genetic evidence reveals temporal change in hybridization patterns in a wild baboon population. *Molecular Ecology*, 17(8), 1998–2011.

Tung, J., Zhou, X., Alberts, S. C., Stephens, M. & Gilad, Y. (2015b). The genetic architecture of gene expression levels in wild baboons. *eLife*, 25, 4.

Turnbaugh, P.J., Ley, R.E., Mahowald, M.A., Magrini, V., Mardis, E.R. & Gordon, J.I. (2006). An obesity-associated gut microbiome with increased capacity for energy harvest. *Nature*, 444(21), 1027–1031.

Turnbaugh, P.J., Ridaura, V.K., Faith, J.J., Rey, F.E., Knight, R. & Gordon, H.A. (2009). The effect of diet on the human gut microbiome, a metagenomic analysis in humanized gnotobiotic mice. *Science Translational Medicine*, 1, 6ra14.

Turner, T.R. (1977). Biological variation in vervet monkeys (Cercopithecus aethiops). PhD dissertation. New York: New York University.

(1981). Blood protein variation in a population of Ethiopian vervet monkeys (*Cercopithecus aethiops aethiops*). *American Journal of Physical Anthropology*, 55, 225–232.

& Weiss, M.L. (1999). Genetics of Old World monkeys: one aspect of the "newer" physical anthropology. In S.C. Strum, D.G. Kindburg & D. Hamburg, eds., *The New Physical Anthropology: Science, Humanism and Critical Reflection*. Upper Saddle River, NJ: Prentice Hall.

Turner, T.R., Anapol, F., & Jolly, C.J. (1997). Growth, development, and sexual dimorphism in vervet monkeys at four sites in Kenya. *American Journal of Physical Anthropology*, 103(1), 19–35.

Turner, T.R., Coetzer, W.G., Schmitt, C.A., Lorenz, J., Freimer, N.B. & Grobler, J.P. (2016). Localized population divergence of vervet monkeys (*Chlorocebus* spp.) in South Africa: evidence from mtDNA. *American Journal of Physical Anthropology*, 159, 17–30.

Turner, T.R., Cramer, J.D., Nisbett, A. & Gray, J. P. (2016). A comparison of adult body size between captive and wild vervet monkeys (*Chlorocebus aethiops sabaeus*) on the island of St. Kitts. *Primates*, 57, 211–220.

Turner, T.R., Hill, R., Coetzer, W.G. & Paterson, L. (2017). A conservation assessment of *Chlorocebus pygerythrus*. In M.F. Child, L. Roxburgh, E. Do Linh San, D. Raimondo & H.T. Davies-Mostert, eds., *The Red List of Mammals of South Africa, Swaziland, and Lesotho, 2016*. Pretoria, South Africa: South African National Biodiversity Institute and Endangered Wildlife Trust.

Turner, T.R., Lorenz, J., Pampush, J.D., Freimer, N. & Grobler, J.P. (2010). The utility of genetic and morphological data in understanding taxonomy in vervet monkeys. Paper presented at the Congress of the International Primatological Society, Kyoto, Japan.

Turner, T.R., Schmitt, C.A., Cramer, J.D., Lorenz, J., Grobler, J.P., Jolly, C.J. & Freimer, N.B. (2018). Morphological variation in the genus *Chlorocebus*: ecogeographic and anthropogenically mediated variation in body mass, postcranial morphology, and growth. *American Journal of Physical Anthropology*, 166(3), 682–707.

Turner, T.R., Whitten, P.L., Jolly, C.J. & Else, J.G. (1987). Pregnancy outcome in free-ranging vervet monkeys (*Cercopithecus aethiops*). *American Journal of Primatology*, 12(2), 197–203.

Turnquist, J.E. & Kessler, M.J. (1989). Free-ranging Cayo Santiago rhesus monkeys (*Macaca mulatta*), I. Body size, proportion, and allometry. *American Journal of Primatology*, 19(1), 1–13.

Uehara, S. & Nishida, T. (1987). Body weights of wild chimpanzees (*Pan troglodytes schweinfurthii*) of the Mahale Mountains National Park, Tanzania. *American Journal of Physical Anthropology*, 72(3), 315–321.

UK10K Consortium, Walter, K., Min, J.L., Huang, J. et al. (2015). The UK10K project identifies rare variants in health and disease. *Nature*, 526, 82–90.

Ulijaszek, S.J. (1993). Evidence for a secular trend in heights and weights of adults in Papua New Guinea. *Annals of Human Biology*, 20(4), 349–355.

van de Waal, E. & Bshary, R. (2010). Contact with human facilities appears to enhance technical skills in wild vervet monkeys (*Chlorocebus aethiops*). *Folia Primatologica*, 81, 282–291.

(2011). Social learning abilities of wild vervet monkeys in a two-step task artificial fruit experiment. *Animal Behaviour*, 81, 433–438.

van de Waal, E., Borgeaud, C. & Whiten, A. (2013a). Potent social learning and conformity shape a wild primate's foraging decisions. *Science*, 340(6131), 483–485.

van de Waal, E., Bshary, R. & Whiten, A. (2014). Field experiments show wild vervet monkey infants acquire maternal food-processing techniques. *Animal Behaviour*, 90, 41–45.

van de Waal, E., Claidière, N. & Whiten, A. (2013b). Social learning and spread of alternative means of opening an artificial food in four groups of vervet monkeys (*Chlorocebus aethiops*). *Animal Behaviour*, 85, 71–76.

(2015). Wild vervet monkeys discriminate and copy alternative methods for opening an artificial fruit. *Animal Cognition*, 18(3), 617–621.

van de Waal, E., Krützen, M., Hula, J., Goudet, J. & Bshary, R. (2012). Similarity in food cleaning techniques within matrilines in wild vervet monkeys. *PLoS One*, 7(4), e35694.

van de Waal, E., Renevey, N., Favre, C.M. & Bshary, R. (2010). Selective attention to philopatric models causes directed social learning in wild vervet monkeys. *Proceedings of the Royal Society of London B*, 277, 2105–2111.

van de Waal, E. & Whiten, A. (2012). Spontaneous emergence, imitation and spread of alternative foraging techniques among groups of vervet monkeys. *PLoS One*, 7 (10), e47008.

van de Woude, S. & Apetrei, C. (2006). Going wild: lessons from T-lymphotropic naturally occurring lentiviruses. *Clinical Microbiology Review*, 19, 728–762.

van der Kuyl, A.C., Kuiken, C.L., Dekker, J.T. & Goudsmit, J. (1995). Phylogeny of African monkeys based upon mitochondrial 12S rRNA sequences. *Journal of Molecular Evolution*, 40(2), 173–180.

van Heuverswyn, F., Li, Y., Neel, C. et al. (2006). Human immunodeficiency viruses, SIV infection in wild gorillas. *Nature*, 444, 164.

van Rensburg, E.J., Engelbrecht, S., Mwenda, J., Laten, J.D., Robson, B.A., Stander, T. & Chege, G.K. (1998). Simian immunodeficiency viruses (SIVs) from eastern and southern Africa, detection of a SIVagm variant from a chacma baboon. *Journal of General Virology*, 79, 1809–1814.

van Schaik, C.P. (1983). Why are diurnal primates living in groups? *Behaviour*, 87, 120–144.

(1989). The ecology of social relationships amongst female primates. In V. Standen & R.A. Foley, eds., *Comparative Socioecology: The Behavioural Ecology of Humans and Other Mammals*. Oxford, UK: Blackwell, pp. 195–218.

van Schaik, C.P. & Kappeler, P.M. (1993). Life history, activity period and lemur social systems. In J. Ganzhorn, ed., *Lemur Social Systems and Their Ecological Basis*. Boston, MA: Springer, pp. 241–260.

van Schaik, C.P. (1996). Social evolution in primates: the role of ecological factors and male behavior. *Proceedings of the British Academy*, 88, 9–31.

van Schaik, C.P. & Hörstermann, M. (1994). Predation risk and the number of males in a primate group, a comparative test. *Behavioral Ecology and Sociobiology*, 35, 261–272.

van Schaik, C.P. & van Hooff, J.A.R.A.M. (1983). On the ultimate causes of primate social systems. *Behaviour*, 85, 91–117.

Van Wegenen, G. & Catchpole, H.R. (1956). Physical growth of the rhesus monkey (*Macaca mulatta*). *American Journal of Physical Anthropology*, 14, 245–273.

Verheyen, W.N. (1962). Contribution to the comparative craniology of primates. The genuses *Colobus* Illeger 1811 and *Cercopithecus* Linne 1758. *Annales du Muse Republique du Congo Belge*, B105, 1–255.

Vigilant, L., Hofreiter, M., Siedel, H. & Boesch, C. (2001). Paternity and relatedness in wild chimpanzee communities. *Proceedings of the National Academy of Sciences of the United States of America*, 98(23), 12890–12895.

Vigue, K. (2008). Behavioral and hormonal variability in vervet monkeys under stressed conditions. MS thesis. Milwaukee, WI: University of Wisconsin-Milwaukee.

Villar, D., Berthelot, C., Aldridge, S. et al. (2015). Enhancer evolution across 20 mammalian species. *Cell*, 160(3), 554–566.

Vinson, A., Prongay, K. & Ferguson, B. (2013). The value of extended pedigrees for next-generation analysis of complex disease in the rhesus macaque. *ILAR Journal*, 54, 91–105.

Vitzthum, V.J. (2008). Evolutionary models of women's reproductive functioning. *Annual Review of Anthropology*, 37, 53–62.

Voruganti, V.S., Jorgensen, M.J., Kaplan, J.R. et al. (2013). Significant genotype by diet (G × D) interaction effects on cardiometabolic responses to a pedigree-wide, dietary challenge in vervet monkeys (*Chlorocebus aethiops sabaeus*). *American Journal of Primatology*, 75, 491–499.

Walker, R., Gurven, M., Hill, K. et al. (2006). Growth rates and life histories in twenty-two small-scale societies. *American Journal of Human Biology*, 18(3), 295–311.

Wallis, J. (2013). Green monkey *Chlorocebus sabaeus*. In R.A. Mittermeier, A.B. Rylands & D.E. Wilson, eds., *Handbook of the Mammals of the World Volume 3: Primates*. Barcelona, Spain: Lynx Edicions, p. 673.

Warren, W.C., Jasinska, A.J., Garcia-Perez, R. et al. (2015). The genome of the vervet (*Chlorocebus aethiops sabaeus*). *Genome Research*, 25(12), 1921–1933.

Washburn, S.L. & DeVore, I. (1961). The social life of baboons. *Scientific American*, 204(6), 62–71.

Watkins, D.I., Burton, D.R., Kallas, E.G., Moore, J.P. & Koff, W.C. (2008). Nonhuman primate models and the failure of the Merck HIV-1 vaccine in humans. *Nature Medicine*, 14, 617–621.

Watts, E.S. & Gavan, J.A. (1982). Postnatal growth of nonhuman primates, the problem of the adolescent spurt. *Human Biology*, 54, 53–70.

Wells, J.C.K., Chomtho, S. & Fewtrell, M.S. (2007). Programming of body composition by early growth and nutrition. *Proceedings of the Nutritional Society*, 66, 423–434.

Weiner, A.S. & Moor-Jankowski, J. (1969). The ABO blood groups of baboons. *American Journal of physical Anthropology*, 30, 117–122.

Weiss, M.L. & Goodman, M. (1971). Genetic structure and systematics of some macaques and men. In A.B. Chiarelli, ed., *Comparative Genetics in Monkeys, Apes, and Man*. London, UK: Academic Press, pp. 129–151.

Whiten, A. (1998). Imitation of the sequential structure of actions by chimpanzees (*Pan troglodytes*). *Journal of Comparative Psychology*, 112, 270–281.

(2005). The second inheritance system of chimpanzees and humans. *Nature*, 437, 52–55.

(2009). The identification of culture in chimpanzees and other animals: from natural history to diffusion experiments. In K.N. Laland & B.G. Galef, eds., *The Question of Animal Culture*. Cambridge, MA: Harvard University Press, pp. 99–124.

Whiten, A. & Mesoudi, A. (2008). Establishing an experimental science of culture, animal social diffusion experiments. *Philosophical Transactions of the Royal Society B*, 363, 3477–3488.

Whiten, A. & van Schaik, C.P. (2007). The evolution of animal "cultures" and social intelligence. *Philosophical Transactions of the Royal Society B*, 362, 603–620.

Whiten, A., Custance, D.M., Gomez, J.C., Teixidor, P. & Bard, K.A. (1996). Imitative learning of artificial fruit processing in children (*Homo sapiens*) and chimpanzees (*Pan troglodytes*). *Journal of Comparative Psychology*, 110, 3–14.

Whitney, A. R., Diehn, M., Popper, S.J. et al. (2003). Individuality and variation in gene expression patterns in human blood. *Proceedings of the National Academy of Sciences of the United States of America*, 100, 1896–1901.

Whittaker, D.J., Morales, J.C., & Melnick, D.J. (2007). Resolution of *Hylobates* phylogeny: congruence of mitochondrial D-loop sequences with molecular, behavioral, and morphological data sets. *Molecular Phylogenetics and Evolution*, 45(2), 620–628.

Whitten, P.L. (1983). Diet and dominance among female vervet monkeys (*Cercopithecus aethiops*). *American Journal of Primatology*, 5, 139–159.

Whitten, P.L. & Turner, T.R. (2004). Male residence and patterning of serum testosterone in vervet monkeys (*Cercopithecus aethiops*). *Behavioral Ecology and Sociobiology*, 56(6), 565–578.

(2008). Ecological and reproductive variance in serum leptin in wild vervet monkeys. *American Journal of Physical Anthropology*, 137(4), 441–448.

(2009). Endocrine mechanisms of primate life history trade-offs: growth and reproductive maturation in vervet monkeys. *American Journal of Human Biology*, 21(6), 754–761.

Wickings, E.J. & Dixson, A.F. (1992). Testicular function, secondary sexual development, and social status in male mandrills (*Mandrillus sphinx*). *Physiology and Behavior*, 52, 909–916.

Widdig, A., Kessler, M.J., Bercovitch, F.B. et al. (2015). Genetic studies on the Cayo Santiago rhesus macaques: a review of 40 years of research. *American Journal of Primatology*, 78(1), 44–62.

et al. (2016a). Genetic studies on the Cayo Santiago rhesus macaques: a review of 40 years of research. *American Journal of Primatology*, 78, 44–62.

Widdig, A., Langos, D. & Kulik, L. (2016b). Sex differences in kin bias at maturation: male rhesus macaques prefer paternal kin prior to natal dispersal. *American Journal of Primatology*, 78, 78–91.

Widdig, A., Nürnberg, P., Krawczak, M., Streich, W.J. & Bercovitch, F.B. (2001). Paternal relatedness and age proximity regulate social relationships among adult female rhesus macaques. *Procdings of the National Academy of Sciences of the United States of America*, 98, 13769–13773.

Widdig, A., Bercovitch, F.B., Streich, W.J., Sauermann, U., Nuernberg, P. & Krawczak, M. (2004). A longitudinal analysis of reproductive skew in male rhesus macaques. *Proceedings of the Royal Society of London*, B271(1541), 819–826.

Wiederholt, R. & Post, E. (2010). Tropical warming and the dynamics of endangered primates. *Biology Letters*, 6, 257–260.

Wiens, J.A. (1976). Population responses to patchy environments. *Annual Review of Ecology, Evolution, and Systematics*, 7, 81–120.

Wilen, R. & Naftolin, F. (1976). Age, weight and weight gain in the individual pubertal female rhesus monkey (*Macaca mulatta*). *Biology of Reproduction*, 15, 356–360.

Willems, E.P. & Hill, R.A. (2009). Predator-specific landscapes of fear and resource distribution, effects on spatial range use. *Ecology*, 90, 546–555.

Wilson, P.J. (1975). The promising primate. *Man*, 5–20.

Wimberger, K., Downs, C.T. & Perrin, M.R. (2010a). Postrelease success of two rehabilitated vervet monkey (*Chlorocebus aethiops*) troops in KwaZulu-Natal, South Africa. *Folia Primatologica*, 81, 96–108.

Wimberger, K., Downs, C.T. & Boyes, R.S. (2010). A survey of wildlife rehabilitation in South Africa: is there a need for improved management? *Animal Welfare*, 19(4), 481.

Wingfield, J.C. (1994). Hormone–behavior interactions and mating systems in male and female birds. In R.V. Short & B.B. Balaban, eds., *The Differences Between the Sexes*. Cambridge: Cambridge University Press, pp. 303–330.

Wingfield, J.C., Hegner, R.E., Dufty, Jr., A.M. & Ball, G.F. (1990). The "challenge hypothesis": theoretical implications for patterns of testosterone secretion, mating systems, and breeding strategies. *The American Naturalist*, 136, 829–846.

Williams, G.C. (1966). *Adaptation and Natural Selection*. Princeton, NJ: Princeton University Press.

Williams-Blangero, S. (1991). Recent trends in genetic research on captive and wild nonhuman primate populations. *Yearbook of Physical Anthropology*, 34, 69–96.

Wilson, M.E., Gordon, T.P., Rudman, C.G. & Tanner, J.M. (1989). Effects of growth hormone on the tempo of sexual maturation in female rhesus monkeys. *Journal of Clinical Endocrinology and Metabolism*, 68(1), 29–38.

Woods, R. P., Fears, S.C., Jorgensen, M.J., Fairbanks, L.A, Toga, A.W. & Freimer, N.B. (2011). A web-based brain atlas of the vervet monkey, *Chlorocebus aethiops*. *NeuroImage*, 54, 1872–1880.

Wrangham, R.W. (1980). An ecological model for the evolution of female-bonded groups of primates. *Behaviour*, 75, 262–300.

(1981). Drinking competition in vervet monkeys. *Animal Behaviour*, 29, 904–910.

Wrangham, R. & Waterman, P. (1981). Feeding behaviour of vervet monkeys on *Acacia tortilis* and *Acacia xanthophloea*, with special reference to reproductive strategies and tannin production. *Journal of Animal Ecology*, 50, 715–731.

Wren, B.T. (2006). Gastrointestinal parasites of the vervet monkey (*Chlorocebus [Cercopithecus] aethiops*) at a Sanctuary in Limpopo Province, South Africa. MS Thesis. Muncie, IN: Ball State University.

(2013). Behavioral ecology of primate–parasite interactions. PhD dissertation. West Lafayette, IN: Purdue University.

Wren, B.T., Gillespie, T.R., Camp, J.W. & Remis, M.J. (2015). Helminths of vervet monkeys, *Chlorocebus aethiops*, from Loskop Dam Nature Reserve, South Africa. *Comparative Parasitology*, 82, 101–108.

Wren, B.T., Remis, M.J., Camp, J.W. & Gillespie, T.R. (2016). Number of grooming partners is associated with hookworm infection in wild vervet monkeys (*Chlorocebus aethiops*). *Folia Primatologica*, 87(3), 168–179.

Wright, E. & Robbins, M.M. (2014). Proximate mechanisms of contest competition among female Bwindi mountain gorillas (*Gorilla beringei beringei*). *Behavioral Ecology and Sociobiology*, 68, 1785–1797.

Xing, J.C., Witherspoon, D.J., Ray, D.A., Batzer, M.A. & Jorde, L.B. (2007). Mobile DNA elements in primate and human evolution. *Yearbook of Physical Anthropology*, 50, 2–19.

Yang, S.H., Cheng, P.H., Banta, H. et al. (2008). Towards a transgenic model of Huntington's disease in a non-human primate. *Nature*, 453, 921–924.

Yatsunenko, T., Rey, F.E., Manary, M.J. et al. (2012). Human gut microbiome viewed across age and geography. *Nature*, 486(7402), 222–227.

Yeager, C.P. (1997). Orangutan rehabilitation in Tannjung Putting National Park, Indonesia. *Conservation Biology*, 11(3), 802–805.

Yildirim, S., Yeoman, C.J., Janga, S.C. et al. (2014). Primate vaginal microbiomes exhibit species specificity without universal *Lactobacillus* dominance. *ISME Journal*, 8, 2431–2444.

Young, C., Majolo, B., Heistermann, M., Schülke, O. & Ostner, J. (2014). Responses to social and environmental stress are attenuated by strong male bonds in wild macaques. *Proceedings of the National Academy of Sciences of the United States of America*, 111, 18195–18200.

Young, C., McFarland, R., Barrett, L. & Henzi, S.P. (2017). Formidable females and the power trajectories of socially integrated male vervet monkeys. *Animal Behaviour*, 125, 61–67.

Zarrei, M., MacDonald, J.R., Merico, D. & Scherer, S.W. (2015). A copy number variation map of the human genome. *Nature Reviews Genetics*, 16, 172–183.

Zhou, X., Wang, B., Pan, Q. et al. (2014). Whole-genome sequencing of the snub-nosed monkey provides insights into folivory and evolutionary history. *Nature Genetics*, 46, 1303–1310.

Zuberbühler, K. (2007). Predation and primate cognitive evolution. In S. Gursky & K.A.I. Nekaris, eds., *Primate Anti-Predator Strategies*. New York: Springer, pp. 3–26.

Zuberbühler, K. & Jenny, D. (2002). Leopard predation and primate evolution. *Journal of Human Evolution*, 43, 873–886.